U. S. MARINE CORPS
SCOUT/SNIPER
TRAINING MANUAL

Desert Publications
El Dorado, AR 71731-1751 U. S. A.

U. S. Marine Corps Scout/Sniper Training Manual

© 1994 by Desert Publications
215 S. Washington Ave.
El Dorado, AR 71730 U. S. A.
800-852-4445
www.deltapress.com

ISBN 0-87947-094-1
10 9 8 7 6 5 4
Printed in U. S. A.

Reprinted Courtesy of Lancer Militaria
Mt. Ida, AR 71957

**Desert Publications is a division of
The DELTA GROUP, Ltd.
Direct all inquiries & orders to the above address.**

Table of Contents

SCOUT/SNIPER INSTRUCTOR
OUTLINES

For Instructional Purposes Only

SCOUT/SNIPER INSTRUCTOR SCHOOL
MARKSMANSHIP TRAINING UNIT
WEAPONS TRAINING BATTALION
MARINE CORPS DEVELOPMENT AND EDUCATION COMMAND
QUANTICO, VIRGINIA 22134

UNITED STATES MARINE CORPS
Marksmanship Training Unit, Weapons Training Battalion
Marine Corps Development and Education Command
Quantico, Virginia

MTU-21

LESSON OUTLINE

SNIPING

INTRODUCTION

1. <u>Gain Attention</u>. How many of you have seen Sands of Iwo Jima?
Remember John Wayne charging up Mt. Surabachi? He's just about to reach
the top when out of a hole in the ground pops Hollywood's infamous
Japanese Sniper. Seconds later, end of John Wayne. Of course, moments
later the Japanese sniper is killed and once again good triumphs over
evil. But again, Hollywood has created an image of a sniper for the
American public. Regretably, that same image is focussed in the minds of
many military people as well. Whenever the word sniper is mentioned,
many of us automatically conjure up images of Germans and Japanese tied
in trees or popping out of holes to shoot John Wayne, Vic Morrow or Tab
Hunter. In reality, nothing could be further from the truth.

2. <u>Motivate</u>. As future company commanders and battalion S-3's, you will
no doubt have snipers in your units and will be responsible for employing
them in an effective manner. So what we've got to do is erase that
Hollywood image and discuss what a sniper really is and what a sniper can
really do.

3. <u>Purpose and Main Ideas</u>. What we are going to do then is to familiarize
you with first, the background of the current sniper program. Second, we
want to discuss the characteristics that you, as a Company C.O., should
look for in a Marine when selecting your potential snipers. Next, we'll
take a look at the current sniper training and organization. Then we'll
examine probably the two most relevent topics for you - the sniper's
equipment and how to employ the sniper and his equipment.

4. <u>Student Objectives</u>. So, at the end of this period of instruction, you
should be able to·

  a. Discuss the background of the present sniper program.

  b. Discuss the desirable characteristics a potential sniper should
possess.

  c. Describe the sniper organization in the infantry regiment and
battalion.

  d. Describe the capabilities and limitations of the sniper's equipment.

  e. Describe the tactical employment of snipers in offensive, defensive
and urban guerrilla/terrorist combat.

BODY

1. <u>General</u>. Before any discussion of snipers or sniping can start, we first must know exactly what a sniper is supposed to be. If there are 150 of you here, I'm sure we've probably got 100 different definitions. But let's discuss the USMC definition of a sniper as contained in FMFM 1-3B. That FMFM states that a sniper is a Marine highly trained in field craft and marksmanship, who delivers long range, precision fire at selected targets from concealed positions in support of combat operations. The key words in this definition are "at selected targets." That's what sets the sniper apart from the ordinary rifleman. Hopefully, any Marine can deliver relatively long range, precision fire from a relatively concealed position. But normally, the average Marine rifleman is not concerned with who or what the target is, as long as it's the enemy. The sniper, on the other hand, is interested in the importance of his target. He engages only those targets which will have a profound influence on the enemy's ability to wage battle. Such targets might include Officers, NCO's, communicators, crew served weapons personnel, observation equipment and sights on howitzers or mortars. Once these types of targets have been identified, then, and only then, will the sniper shoot that extremely accurate, one round.

2. <u>Background</u>. But before we examine the sniper himself any closer, let's examine what has brought the sniper back into the Marine Corps inventory.

    a. CMC Project 30-71-04, Infantry Organization and Weapons Systems, 1973-1977, identifies requirement for snipers because

        (1) The M16A1 does not possess the inherent accuracy to consistently engage and hit targets at ranges greater than 500 meters.

        (2) At these longer ranges, the skill necessary to hit a point target of approximately man size is considerable and not possessed by the average Marine rifleman.

        (3) Combat situations of all levels may require the <u>selective</u> engagement of targets at ranges greater than 500 meters.

    b. The study recommended:

        (1) Re-introduction of MOS 8541 Scout-Sniper

        (2) Introduction of eight (8) sniper teams per infantry battalion with such duties performed on a collateral basis.

        (3) Training be conducted in conjunction with the Competition-In-Arms Program and the MTU, WTBN, MCDEC, Quantico, Virginia.

    c. CMC Message 301417E Mar 1977 implemented the recommendations contained in the study mentioned above. Since that time, some changes have been made and more changes are being considered, but what we're going to discuss is the program as it exists right now.

3. Selection of Personnel

   a. General. Candidates must be carefully screened by unit commanders.

      (1) Requires high degree of motivation.

      (2) Ability to learn variety of skills.

   b. Selection Requirements

      (1) Marksmanship

         (a) Repeated annual qualification as expert is most desirable.

         (b) Successful participation in annual Competition-In-Arms Program.

         (c) Preferably an extensive hunting background.

      (2) Physical Condition

         (a) Performance on physical fitness test.

         (b) Participation in athletics, particularly team sports.

         (c) Undesirable Physical Characteristics

            1 Glasses - cause glare or might be broken

            2 Smoking - smoke can be seen and smelled and abstinence might cause needless shaking.

            3 Left-Handed - requires needless motion to operate the bolt.

      (3) Mental Condition

         (a) Must not be susceptible to anxiety or remorse.

         (b) Must display effective decisiveness, self reliance, good judgement and common sense.

         (c) Must display composure under stress.

      (4) Field Craft

         (a) Often a background in hunting, trapping, forestry, or Boy Scouting will identify potential snipers.

         (b) Must know and understand principles of individual movement, camouflage, etc...

4. **Sniper Training**

   a. **Division Level Sniper School**

      (1) Purpose: To teach basic sniper skills and employment techniques.

      (2) Training Objectives; ensure the student is capable of:

         (a) One shot kill on stationary target out to 1,000 yards.

         (b) One shot kill on moving target out to 800 yards.

         (c) Stalking undetected for 1,000 yards.

         (d) Judging distance to 1,000 yards with ± 10% error.

         (e) Using maps and aerial photos to select routes of advance and positions.

         (f) Observing, collecting, and reporting intelligence data.

         (g) Selecting and preparing sniper positions.

         (h) Employment of a sniper team in combat.

   b. **Marine Corps Level Scout-Sniper Instructor School**

      (1) Purpose: To refine the basic sniper skills and teach advanced employment techniques.

      (2) Training Objectives; Ensure the student is capable of:

         (a) Teaching basic sniper skills in Division Level Schools.

         (b) Leading and employing a Battalion Level Sniper Section.

5. **Sniper Organization**

   a. **Rifle Company, Infantry Battalion   T/O 1037M**

      (1)
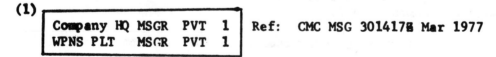

| Company HQ | MSGR | PVT | 1 |
| WPNS PLT | MSGR | PVT | 1 |

      Ref: CMC MSG 301417Z Mar 1977

      (2) Billets

         (a) Collateral duty

         (b) Both qualified Scout-Snipers (MOS 8541)

      (3) Weapons: One (1) M40A1 Sniper Rifle per team.

b. <u>H & S Company, Infantry Battalion T/O 1037M</u>

(1)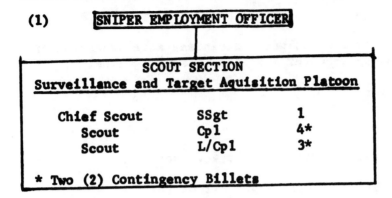

(2) Billets

    (a) Collateral duty
    (b) All qualified Scout-Snipers (MOS 8541)

(3) Weapons: Four (4) M40A1 Sniper Rifles

c. <u>HQ Company, Infantry Regiment T/O 1096M</u>

Scout-Sniper Platoon

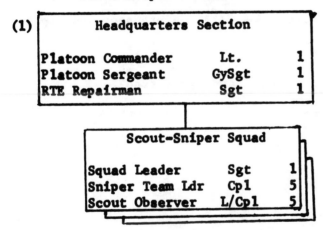

(2) Billets: Cadred at zero level, except in war or contingency situations.

(3) Weapons: Five (5) M40A1 Sniper Rifles per squad.

6. Sniper Equipment

    a. M40A1 Sniper Rifle

        (1) New modifications include· stainless steel barrels, camouflaged fiberglass stocks, glassed receiver and steel trigger guards.

        (2) Capabilities.

            (a) Maximum effective range: 1,000 meters

            (b) Less than minute of angle accuracy

            (c) 3X - 9X telescopic sight allows for·

                1 Improved target identification and location

                2 Use at night under artificial illumination

        (3) Limitations.

            (a) Bolt action; not capable of semi-automatic fire

            (b) Must hold off for wind and elevation

            (c) Requires special (MOS 2112) maintenance and repair

        (4) Ammunition; 7.62mm

            (a) M118 Lake City Match Ammo (preferred).

            (b) M80 Ball Ammo

            (c) M61 Armor Piercing

            (d) M62 Tracer

    b. M17A1 Binoculars

        (1) 7X50 wide angle view

        (2) Used for spotting targets and estimating distance.

    c. M49 Spotting Scope

        (1) 20 power, for detailed observation

        (2) Used for spotting targets, calculating wind direction and velocity and calling shots.

d. **Camouflage Uniform/Ghillie Suit**

    (1) It is a loose fitting camouflage jacket or smock with colored strips of burlap garnish attached.

    (2) A face veil is attached to assist in breaking up the outline of the sniper's head and shoulders.

    (3) It is an extremely effective form of personal camouflage when used in conjunction with existing natural camouflage materials.

e. **AN/PVS 2B Starlight Scope**

    (1) Can be mounted on either M14 or M16; <u>not M40A1</u>.

    (2) Maximum effective range - 300 yards.

f. **.50 Caliber Machine Gun**

    (1) Used effectively by Marine Corps snipers in Vietnam.

    (2) 12X to 20X scopes allowed for kills at ranges approaching 3,000 yards.

    (3) Used primarily in defensive, static positions.

7. **Sniper Employment**

a. **General.**

    (1) The best results from sniping are only achieved if the employment of snipers is carefully organized. For this reason, except for specific purposes, the planning should be done no lower than Company level.

    (2) Effective sniping by well-trained and well-organized snipers will do more than inflict casualties and cause inconvenience to the enemy. It will have a marked effect on the security and morale of the troops and will result in a constant source of information about the enemy.

    (3) The method by which snipers are employed will be governed by many factors, such as: the nature of the ground, the distance between forward troops and the enemy, the existence of obstacles, the degree of initiative shown by the enemy and the number of snipers available.

    (4) Snipers are largely a weapon of opportunity and the broadest principles only can be given as to their employment. These principles must be applied with imagination and be guided by a commander's sound grasp of the sniper team's capabilities.

    (5) The attachment of the sniper team to any unit provides that unit with an additional supporting arm. It allows the unit to engage point targets at distances beyond the range of the service rifle.

(6) The sniper can operate at night as effectively as he can in the day.

(a) The M40A1 with the standard Redfield 3x-9x scope gathers sufficient light from artificial illumination to enable the sniper to engage targets out to 600 yards.

(b) The AN/PVS-2A allows the sniper to engage targets out to approximately 300 yards.

b. Offensive Combat

(1) General. Support the infantry by delivering precision long-range fire at:

(a) Automatic weapons
(b) Artillery forward observers
(c) Enemy personnel (Particularly officers and NCO's)
(d) Optical devices used for observation
(e) Fleeing enemy personnel

And by:

(f) Protecting the flanks of attacking units
(g) Covering by fire, gaps between attacking elements
(h) Participating in repelling counterattacks
(i) Eliminating enemy snipers left to delay an advance.

(2) Frontal Attack

(a) Direct fire at exposed enemy troops
(b) Deliver fire into apertures of enemy bunkers
(c) Destroy enemy crew served weapons and crews
(d) Deliver long-range fire at targets located beyond the objective when maneuvering infantry mask the fire on the objective.
(e) Provide fire to protect flanks or at small, isolated, resistance pockets which have been bypassed.
(f) Fire at targets threatening a counterattack or fire at fleeting enemy personnel.
(g) Fire at selected targets just prior to H-Hour.

(3) Tank-Infantry Attack

(a) Mark targets for tanks at ranges up to 900 meters by use of tracer rounds.
(b) Provide protection for tanks by destroying concealed anti-tank weapons and crews.

(4) Mechanized Infantry Attack

(a) Sniper team support is limited to periods when the armored group is halted.
(b) Provide long-range observation and long-range support fire.

(5) **Attack on Fortified Areas**

(a) Deliver precision fire into observation posts embrasures and at exposed personnel.

(b) Targets are engaged selectively to insure systematic reductions of the enemy's defenses through destruction of his mutual support capabilities.

(6) **Attack on a Built-Up Area**

(1) Operate at a sufficient distance from advancing infantry to keep from becoming involved in fire fights, but close enough to kill more distant targets which threaten the advance.

(2) Some sniper teams should operate independently of the infantry to search for targets of opportunity and particularly for enemy snipers.

(7) **River Crossing Operations**

(a) Assume a position where the entire width of the crossing area can be covered prior to the crossing with the primary mission of observing.

(b) Support the crossing by observation and suppression of the enemy's observation and fire.

(c) During the actual crossing, snipers hold their fire, they preserve secrecy unless targets appear which threaten the operation.

(8) **Patrolling**

(a) **Reconnaissance Patrol**·

1 Snipers normally remain with the security element to provide long-range protection for the recon element.

2 Normally, the only appropriate time to fire at a target of opportunity is when departure from position is imminent and firing will not endanger the success of the patrol.

(b) **Combat Patrol**

1 **Raids**

a If maximum firepower is essential, or the size of the patrol is limited, snipers may not be included.

b A sniper team should be attached to the security element.

c Assist in observing, preventing enemy escape and covering the withdrawal of the assault force.

    <u>d</u> Upon withdrawal from the rallying point, the sniper team may be left behind to delay and harrass the enemy pursuits.

    <u>2</u> Ambush Patrol

     <u>a</u> The sniper teams should be located at both ends of the ambush, away from the main body.

     <u>b</u> The sniper seeks leaders, radio operators and crew served weapons personnel as primary targets.

     <u>c</u> If the enemy is mounted, every effort should be made to kill the drivers of the lead and rear vehicles to block the road, prevent escape and create confusion.

     <u>d</u> The sniper may be retained in position long enough to cover the withdrawal of the ambush unit.

  (9) <u>Extended Daylight Ambush</u>

    (a) It is a long-range ambush conducted exclusively by snipers from preselected, concealed positions in areas where there is likelihood of encountering the enemy.

    (b) It is employed to isolate areas within the battle area by restricting enemy movement, to create fear and confusion among enemy troops and to gain information.

  (10) <u>Helicopter Insertion:</u> Helicopters may be used to insert sniper teams into areas of operation when:

    (a) The selected sniper position is located an excessive distance from friendly lines.

    (b) The situation requires immediate employment.

    (c) The route to a selected sniper position is unduly difficult or heavily saturated with the enemy.

    (d) It is desired to create an adverse psychological effect on the enemy by killing key personnel deep in enemy controlled areas.

    (e) There is a requirement for sniper firing to support a helicopterborne assault or to secure terrain around a landing zone.

    (f) Required as diversionary actions or in response to ambushes of friendly forces.

c. Defensive Combat

    (1) General

        (a) The commander of the unit to which snipers are attached decides the degree to which snipers will participate in the defense.

        (b) Snipers should submit recommendations to the unit commander on employment and positions after taking into consideration:

            <u>1</u> Terrain
            <u>2</u> Security
            <u>3</u> Mutual Support
            <u>4</u> Primary and Supplementary Positions
            <u>5</u> Coordinated Barrier Planning
            <u>6</u> Rate of Fire

    (2) Outposts

        (a) Snipers are assigned to provide long-range precision fire to cause premature deployment of advancing enemy forces and to augment the outposts observation capabilities.

        (b) Snipers can be deployed to cover the withdrawal of outposts by delaying the enemy advance.

    (3) Security. Sniper teams may be assigned any task involving local security <u>during the daylight hours</u> and enhance flank security by providing long-range observation and precision fire.

    (4) Area Defense

        (a) Forward elements The sniper team is assigned the responsibility of defending critical avenues of approach and for firing at targets of opportunity. (If possible, from positions slightly to the rear of the of the frontline rifle companies.)

        (b) Reserve elements. A sniper team's observation capabilities are ideally suited for assignment to security and surveillance missions with the reserve unit. When assigned, the sniper team can:

            <u>1</u> Support the FEBA units by fire
            <u>2</u> Protect key terrain features in the rear
            <u>3</u> Control the most dangerous approaches through the battle area.

    (5) Perimeter defense. The sniper team is positioned on rising terrain near the center of the perimeter, providing the position affords all around observation of avenues of approach and good fields of fire. Maximum emphasis is placed upon mutual support within the perimeter.

    (6) Reverse Slope defense. The sniper team is positioned with the security group on or just forward of the topographical crest to provide long-range support and observation.

(7) **Defense of built-up areas.** Snipers are preferably positioned in buildings of masonry construction which offer the best long-range fields of fire and all around observation. The sniper's mission may include:

      (a) Counter-sniper fire

      (b) Firing at targets of opportunity

      (c) Denying the enemy access to certain areas or avenues of approach.

      (d) Provide fire support over barriers and obstacles

      (e) Surveillance of the flanks and rear areas

      (f) Supporting counter-attacks

      (g) Prevention of enemy observation

(8) **Defense of a river line**

      (a) Snipers are employed initially with the covering force which remains on the enemy side of the river to harrass him and to delay his advance.

      (b) When forced to retire, the sniper should be located as far above or below possible fording sites consistent with observation and fields of fire.

(9) **Mobile defense.** Snipers should be assigned to any size unit assigned the mission of establishing strong points. If a sufficient number of sniper teams are available, they should be assigned to the strong point reserves to cover withdrawal of the strong point.

(10) **Retrograde operations.** Snipers are assigned missions of supporting the action by delaying and inflicting casualties upon the enemy; observation, covering avenues of approach and obstacles by fire, harrassing the enemy by causing him to prematurely deploy and if the situation permits, directing artillery fire on large groups of the enemy.

d. **Urban Guerrilla Warfare**

(1) **General.** The role of the sniper in an urban guerrilla environment is to dominate the area of operations by delivery of selective, aimed fire against **specific** targets as authorized by local commanders. Usually, this authorization only comes when such targets are about to employ firearms or other lethal weapons against the peace keeping force or innocent civilians. The sniper's other role, and almost equally important, is the gathering and reporting of intelligence.

(2) **Tasks.** Within the above role, some specific tasks which may be assigned include:

      (a) When authorized by local commanders, engaging dissidents/urban guerrillas when involved in hijacking, kidnapping, holding hostages, etc.

      (b) Engaging urban guerrilla snipers as opportunity targets or as part of a deliberate clearance operation.

      (c) Covertly occupying concealed positions to observe selected areas.

(d) Recording and reporting all suspicious activity in the area of observation.

(e) Assisting in coordinating the activities of other elements by taking advantage of hidden observation posts.

(f) Providing protection for other elements of the peace keeping force, including firemen, repair crews, etc.

(3) Limitations: In urban guerrilla operations there are several limiting factors that snipers would not encounter in a conventional war:

(a) There is no FEBA and therefore no "No Mans Land" in which to operate. Snipers can therefore expect to operate in entirely hostile surroundings in most circumstances.

(b) The enemy is covert, perfectly camouflaged among and totally indistinguishable from the everyday populace that surrounds him.

(c) In areas where confrontation between peace keeping forces and the urban guerrillas takes place, the guerrilla dominates the ground entirely from the point of view of continued presence and observation. Every yard of ground is known to them; it is ground of their own choosing. Anything approximating a conventional stalk to and occupation of, a hide is doomed to failure.

(d) Although the sniper is not subject to the same difficult conditions as he is in conventional war, he is subject to other pressures. These include not only legal and political restraints, but also the requirement to kill or wound without the motivational stimulus normally associated with the battlefield.

(e) Normally in conventional war, the sniper needs no clearance to fire his shot. In urban guerrilla warfare, the sniper must make every effort possible to determine in each case the need to open fire and that it constitutes reasonable/minimum force under circumstances.

(4) Methods of Employment

(a) Sniper Cordons/Periphery OP's

1 The difficulties to be overcome in placing snipers in heavily populated, hostile areas and for them to remain undetected, are considerable. It is not impossible, but it requires a high degree of training, not only on the part of the snipers involved, but also of the supporting troops.

2 To overcome the difficulties of detection and to maintain security during every day sniping operations, the aim should be to confuse the enemy. The peace keeping forces are greatly helped by the fact that most "trouble areas" are relatively small, usually not more than a few hundred yards in dimension. All can be largely dominated by a considerable number of carefully sited OP's around their peripheries.

       _3_ The urban guerrilla intelligence network will eventually establish the locations of the various OP's. By constantly changing the OP's which are in current use it is impossible for the terrorist to know exactly which are occupied. However, the areas to be covered by the OP's remain fairly constant and the coordination of arcs of fire and observation must be controlled at a high level, usually battalion. It may be delegated to company level for specific operations.

       _4_ The number of OP's required to successfully cordon an area is considerable. Hence, the difficulties of sustaining such an operation over a protracted period in the same area should not be under-estimated.

    (b) <u>Sniper Ambush</u>

       _1_ In cases where intelligence is forth coming that a target will be in a specific place at a specific time, a sniper ambush is frequently a better alternative than a more cumbersome cordon operation.

       _2_ Close reconnaissance is easier than in normal operation as it can be carried out by the sniper as part of a normal patrol without raising any undue suspicion. The principal difficulty is getting the ambush party to its hide undetected. To place snipers in position undetected will require some form of a deception plan. This often takes the form of a routine search operation in at least platoon strength. During the course of the search, the snipers position themselves in their hide. They remain in position when the remainder of the force withdraws. This tactic is especially effective when carried out at night.

       _3_ Once in position the snipers must be prepared to remain for lengthy periods in the closest proximity to the enemy and their sympath-izers.

       _4_ Their security is tenuous at best. Most urban OP's have "dead spots" and this combined with the fact that special ambush positions are frequently out of direct observation by other friendly forces makes them highly susceptible to attack, especially from guerrillas armed with explosives. The uncertainty about being observed on entry is a constant worry to the snipers. It can and does have a most disquieting effect on the sniper and underlines the need for highly trained men of stable character.

       _5_ If the ambush position cannot be directly supported from a permanent position, a back up force must be placed at immediate notice to extract the snipers after the ambush or in the event of compromise. Normally it must be assumed that after the ambush, the snipers cannot make their exit without assistance. They will be surrounded by large, extremely hostile crowds, consequently the "back-up" force must not only be close at hand, but also sufficient in size.

OPPORTUNITY FOR QUESTIONS AND COMMENTS

SUMMARY

1. <u>Re-Emphasize</u>. During this period of instruction, we discussed the sniper program as it stands today. We've examined what characteristics you should look for in potential snipers. We've looked at sniper training at both the Division and Marine Corps Level. We also discussed the current sniper organization and equipment. Finally, we briefly set down some basic sniper employment guidelines.

2. <u>Re-Motivate</u>. I hope that when you're back out in the FMF and you see Marines wearing Ghillie Suits and carrying sniper rifles, you will erase the Hollywood image from your mind and think of them only as a highly skilled, highly trained, extremely effective supporting arm.

26468 MCDEC, QUANTICO, VA.

UNITED STATES MARINE CORPS
SCOUT/SNIPER INSTRUCTOR SCHOOL
Marksmanship Training Unit, Weapons Training Battalion
Marine Corps Development and Education Command
Quantico, Virginia 22134

SI 0001
L, D, A

_____
DATE

CARE AND CLEANING OF THE M40A1 SNIPER RIFLE
AND EQUIPMENT

Detailed Outline

INTRODUCTION   (60 Min.)

1. **Gain Attention.** (HOLD UP SNIPER RIFLE) This is the M40A1 Sniper Rifle, the first Marine-proof weapon ever developed. It will withstand rain, sleet, wind, snow, gloom of night and the abuse which any weapon receives from the average Marine. However, it will not function properly if it is not cleaned regularly and correctly.

2. **Motivate.** A sniper is one who, among other things, delivers precision fire on selected targets. If he has cared for his weapon properly, he will achieve a hit from as far as 1000 yards or more. If he has not cleaned his weapon well, it will lose it's ability to fire accurately, and the sniper will become just another rifleman who has to lug around a heavy, single shot rifle.

3. **State Purpose and Main Ideas.**

    a. **Purpose.** To make the student aware of the various types of rifle and telescope cleaning equipment and how to use them.
    b. **Main Ideas.** Describe and demonstrate the use of the brass cleaning rod, bore and chamber brushes, lens brushes, solvents, and lubricants as used with the M40A1 sniper rifle and optical equipment.

4. **Learning Objectives.** Upon completion of this period of instruction, the student will be able to:
    a. Clean and care for the M40A1 rifle in differing climates, and
    b. Clean and care for the optical equipment used with the M40A1.

TRANSITION. Cleaning weapons is something you have all done from time to time since you've joined the Marine Corps, but passing inspection was usually the only measure of how well you had done your job. How many of you have considered that you might have damaged your weapon by cleaning it improperly?

BODY.

1. The types of equipment used in cleaning the rifle and it's optical accessories include the following:
   a. 7.62mm Bore Brush
   b. .45 cal. Bore Brush
   c. A brass cleaning rod with handle and patch holder
   d. LSA or LAW, as required
   e. Cleaning Solvent
   f. Bore Cleaner
   g. Lens Paper
   h. Camel Hair Brush
   k. .30 cal./7.62mm/.308 cal. cleaning patches

2. To preclude damage, the following precaution will be observed prior to, and while cleaning the rifle:
   a. Assemble the brass cleaning rod with a bore brush on the end.
   b. Lay the rifle on a cleaning table or other flat surface with the muzzle away from the body and the sling down. Make sure you do not strike the muzzle or the sight on the table while doing this.
   c. Always clean the bore from the chamber toward the muzzle.
      (1) With the bore brush, push the brush all the way through until it protrudes from the muzzle and then pull it back until it clears the chamber.
      (2) With the patch, push it all the way through until it protrudes from the muzzle , take the patch off the tip, and then pull the rod back through the chamber.
      (3) While cleaning, keep the muzzle depressed slightly lower than the chamber. This prevents bore cleaner from running into the receiver or firing mechanism.
      (4) Run the bore brush through the bore a minimum of twenty times.
      (5) Using a section of the cleaning rod and the .45 cal. cleaning brush dipped in bore cleaner, clean the chamber by rotating the brush eight to ten times. Do not scrub the brush in and out of the chamber.
      (6) Reassemble the cleaning rod with a swab holder attached. Insert a clean patch and run the patch through the bore. Change patches and continue until a patch comes out clean.

      (7) Finally, with a clean patch, apply a very light coat of LSA to the bore and chamber.

3. Observing the above precautions, clean the rifle and sight in the following manner:
   a. Before Firing. Clean and dry the bore with cleaning rod and patches. Clean and coat all external metal surfaces with a light coat of LSA. Lubricate the bolt with LSA. Clean the sight's lens with lens brush and tissue.
   b. During Firing and While in the Field.
      (1) Stock. Keep free of mud and other debris. Keep foreign objects from between the stock and barrel by sliding a cloth under the barrel and drawing back and forth in a "shoe shine" fashion. Never attempt to remove the torque screws that hold the receiver into the stock.
      (2) Sight. Keep metal surfaces clean and lightly coated with LSA. Keep the lens clean with lens brush and tissue. Keep lens covers on whenever possible. DO NOT touch lenses with fingers or with LSA.

(3) <u>Rifle</u>. Keep the bore, chamber and receiver clean and free of obstructions. Again, <u>NEVER attempt to remove the torque screws that hold the receiver into the stock!!</u>

(4) <u>Ammunition</u>. Keep dry and wiped clean.

c. <u>After Firing</u>. Clean the rifle immediately after firing and for three days thereafter. This cleaning should include the stock, the sight, all external metal surfaces and the bolt, chamber and bore as follows:

(1) Clean the stock first with a clean cloth slightly dampened with clean water. Then wipe it dry.

(2) Clean the sight next with a lens brush and tissue. Replace the protective covers and lightly coat the metal surfaces with LSA.

(3) Remove the bolt and using a cloth with cleaning solvent or bore cleaner, wipe all outer surfaces, paying particular attention to the face of the bolt.

(4) Run the bore brush through the bore a minimum of twenty times.

(5) Using the .45 cal. cleaning brush dipped in bore cleaner, clean the chamber by rotating the brush eight to ten times. <u>DO NOT</u> scrub the brush in and out of the chamber.

(6) Reassemble the cleaning rod with a patch holder attached. Insert a clean patch and run it through the bore. Change patches and continue until a patch comes out clean.

(7) With the cleaning rod and patch, reapply a generous coat of bore cleaner in the bore and chamber and let it remain overnight. Be sure that the muzzle remains lower than the chamber, so that bore cleaner will not run into the receiver.

(8) The next day, punch out the bore and repeat the process for three consecutive days. At the end of the third day, remove all traces of bore cleaner from the bore and chamber area and apply a very light coat of LSA.

4. <u>Care and cleaning under unusual conditions</u>. The instructions already provided are for normal temperatures and normal conditions. Modifications must be made for other than normal conditions. The following information is therefore provided as basic guidance.

    a. <u>Desert Operations</u>.
        (1) Use less lubricant and preservative oil (LSA)
        (2) Keep rifle free of sand by use of the carrying case.
        (3) Keep the sight protected from the direct rays of the sun.
        (4) Keep ammunition clean and protected from the direct rays of the sun.
        (5) Use a toothbrush to remove sand from the bolt and receiver.
        (6) Clean the chamber and bore daily.
        (7) Protect the muzzle and receiver from blowing sand by covering with a clean cloth.
        (8) Keep the lens caps on the sight when not in use.
    b. <u>Arctic Operations (Extreme Cold)</u>.
        (1) Use less lubricant and preservative.
        (2) Substitute LAW for LSA.
        (3) Keep the rifle in it's carrying case when not in use.
        (4) Keep ammunition dry, clean and at the same temperature as the rifle.
        (5) Store rifle in cold protected place to prevent the rifle from sweating or the sight from fogging when taken outside.
        (6) Do not breathe on the sight's lens.
        (7) Keep lens covers on when not in use.
        (8) Check bore for snow or ice obstructions before firing the first shot.
        (9) Use a clean cloth to protect the muzzle and receiver from blowing snow.

c. Jungle Operations (High Humidity).
    (1) Use more lubricant or LSA
    (2) Keep rifle in the carrying case when not in use
    (3) Protect from rain whenever possible
    (4) Keep ammunition clean and dry
    (5) Clean rifle and it's bore and chamber daily and preserve with LSA.
    (6) Request a sight replacement from the supporting maintenance unit when moisture or fungus can be seen inside the lens.
    (7) Clean and dry the stock daily
    (8) Dry the carrying case and the rifle in the sun whenever possible.

d. Beach and Ocean Areas (Sand and Salt Water). Use the same procedure as set forth for jungle operations, except where sand is a problem. Then use a toothbrush to remove sand from the receiver area and the lens brush and the tissues to clean the sight's lens. Cover the muzzle and receiver with a clean cloth to protect from blowing sand.

OPPORTUNITY FOR QUESTIONS AND COMMENTS   (1 Min.)

SUMMARY

1. Reemphasize. We have just looked at the equipment necessary to clean the sniper rifle and it's optical accessories, and we have seen the correct way to clean and care for these things under all different climatic and weather conditions.

2. Remotivate. The M40A1 is a rifle which was built by experienced armorers, to exacting specifications. It is the finest sniper rifle in the world today. Improper care or lack of proper cleaning, however, can render this rifle as useless as an M16 for sniping purposes.

UNITED STATES MARINE CORPS
SCOUT/SNIPER INSTRUCTOR SCHOOL
Marksmanship Training Unit, Weapons Training Battalion
Marine Corps Development and Education Command
Quantico, Virginia 22134

SI0002
L, D, A

_____
(DATE)

SHOOTING POSITIONS
AND
BOLT OPERATION
TITLE

DETAILED OUTLINE

INTRODUCTION

1. **Gain Attention.** Throughout position training, the sniper must be continually checked on the proper application of position principles. This check is the responsibility of you, the instructor, who must closely observe the sniper during all phases of fundamental training.

2. **Motivate.** On the battlefield, a sniper must assume the steadiest possible position which can allow observation of the target area and provide some cover and/or concealment.

3. **State Purpose and Main Ideas.**

   a. **Purpose.** The purpose of this period of instruction is to make the student aware of the various types of positions used by the sniper and the proper steps used to operate the bolt.

   b. **Main Ideas.** The main ideas to be discussed are the following:

      (1) Factors common to all positions
      (2) Positions
      (3) Supported Positions
      (4) Bolt Operation

4. **Learning Objectives.** Upon completion of this period of instruction, the student will:

   a. Demonstrate all positions both supported and unsupported.

   b. Demonstrate tne proper method of working the bolt.

TRANSITION. Considering the many variables of terrain, vegetation, and tactical situations, there are innumerable possible positions, however, in most instances, they will be variations of the five to be discussed.

BODY

1. __Factors Common to All Positions.__ There are seven factors involved in holding the rifle while aiming and firing. These factors are the same for all firing positions, however, the precise manner in which they apply differs slightly with the various positions.

    a. __Left Hand and Elbow.__ The left hand if forward with the palm of the hand against the upper sling swivel when possible. (TA) The wrist is straight and locked so that the rifle rests across the heel of the hand. The hand itself is relaxed. The fingers can be curled against but not gripping the stock since the rifle should rest on the left hand. The left elbow should be directly under the receiver of the rifle, or as close to this position as the conformation of the body will permit. With the left elbow directly under the rifle, the bones (rather than the muscles) support the rifle's weight. (TA) The farther away from this position that the elbow is located, the greater will be the muscular effort necessary to support the rifle. The resulting muscle tension causes trembling and a corresponding movement of the rifle. The sniper trainee, by trial and error, must find the left hand and elbow position best suited to him to avoid tension and trembling.

    b. __Rifle Butt in The Pocket of The Shoulder.__ The sniper places the rifle butt firmly into the pocket of the right shoulder. The proper placement of the butt helps steady the rifle, prevents the rifle butt from slipping on the shoulder during firing, and lessens the effect of recoil.

    c. __Grip of the Right Hand.__ The right hand grips the small of the stock firmly but not rigidly. A firm rearward pressure is exerted by the right hand to keep the rifle butt in it's proper position in the pocket of the shoulder, and to keep the butt secure enough against the shoulder to reduce the effects of recoil. The thumb extends over the small of the stock to enable the sniper to obtain a stock weld. The trigger finger is positioned on the trigger so there is no contact between the finger and the side of the stock. This permits the trigger to be pressed straight to the rear without disturbing the aim.

    d. __Right Elbow.__ The placement of the right elbow gives balance to the sniper's position. Correctly positioned, the elbow helps form the shoulder pocket. The exact location of the right elbow varies with each position.

    e. __Stock Weld.__ A stock weld is when the cheek is placed directly on the rifle stock in the same place each and every time. The firm contact between the cheek and the rifle stock enables the head and weapon to recoil as one unit, thereby facilitating rapid recovery between rounds. The sniper must consciously grip the rifle the same way each and every time so the stock weld will be the same each and every time, enabling the eye to be positioned the same distance behind the eyepiece (eye relief) of the telescope each time the rifle is aimed and fired. This guarantees the same field of vision with each sight picture, thus further assisting in accurate aiming.

f. <u>Breathing</u>. Normal breathing, while aiming and firing the rifle, will cause a movement of the rifle. To avoid this, the sniper must learn to hold his breath for the few seconds required to aim and fire. He takes a normal breath, releases part of it, and holds the remainder. He should not hold his breath for more than approximately 10 seconds or his vision may blur and lung strain may cause muscular tension.

g. <u>Relaxation</u>. The sniper must relax properly in each firing position. Undue strain indicates that muscles are doing work which should be done by bones and that the sniper has, therefore adopted an improper position. He must adjust his position until he is able to relax without violating any of the other holding rules. In a proper firing position, the sniper can relax, but still maintain his sight picture.

2. <u>Positions</u>

a. <u>Prone Position</u>. (TA) The prone position is a relatively steady position which is easy to assume. The position presents a low silhouette and is well adapted to the principle of cover and support. To assume the prone position, the sniper stands facing the target with the left hand forward to the upper sling swivel and the right hand grasping the stock at the heel of the butt. He spreads his feet comfortably apart, shifts his weight slightly rearward and drops to his knees. He leans forward and places his right hand well forward on a line between his right knee and the target and rolls down on his left side, placing his left elbow well forward on the same line. (The rifle is grounded gently to protect the telescopic sight.) With his right hand, he places the butt of the rifle into his right shoulder. He then grips the small of the stock with his right hand and lowers his right elbow to the ground so that his shoulders are approximately level. He then secures a stock weld and relaxes into the tension of the sling. To adjust the natural point of aim to the target, he uses his left elbow as a pivot to move his body until the crosshairs are on the target. He has a well-balanced position if the crosshairs move between 6 o'clock and 12 o'clock, as he breathes. (TA)

b. <u>Hawkins Position</u>. (TA) When firing from a very low bank or a fold in the ground, great steadiness and excellent concealment may be obtained by using this position. The sling is not used and the left hand is positioned forming a fist and grasping the upper sling swivel. (TA) The left arm is straight and the forearm should, if possible, also be in contact with the forestock. The butt of the rifle is on the ground under the right armpit. To control recoil on firing, the left hand must maintain a forward pressure; the left arm should be locked so that recoil is felt in the left shoulder. The Hawkins position is the steadiest position of all and the firer presents a very low silhouette. By holding the rifle by the upper sling swivel with the left hand, the resulting mean point of impact will be practically identical to that of the prone supported position. When the ground is very soft or when using certain slopes, it may not be possible to obtain sufficient depression of the muzzle to use the normal Hawkins position. In the modified position, the butt is placed in the shoulder or on the upper arm. (TA)

c. **Sitting Position.** There are three variations of the sitting position: open legs, crossed legs, and crossed ankle positions. The positions are equally satisfactory, depending on the sniper's body conformation. He must choose the one which gives him the most stability and ease.

(1) **Open Leg.** (TA) For the open leg position, the sling is shortened about 2 or 3 inches from the prone position adjustment. The sniper then faces right from the target, crosses the left foot over the right foot, and sits down. He extends his legs a comfortable distance and spreads his feet up to 36 inches apart. Bending forward at the waist, the sniper aligns his left upper arm against the inside of his left shin. With the right hand at the butt, he pushes the rifle forward and places the butt into the right shoulder. He then moves the right hand forward, grasps the small of the stock, and lowers the upper arm until it rests inside the right knee. By pointing his toes inward, he prevents his knees from spreading and maintains pressure on his upper arms. The position is completed by relaxing the weight forward and assuming the correct stock weld.

(2) **Crossed Leg.** (TA) The difference between the crossed leg and the open leg positions is very slight. For the crossed leg position, the sniper proceeds as for the open leg position except that after sitting down, he simply keeps his feet in place and positions his upper arms inside his knees. Many snipers use the crossed leg position because it can be assumed quickly.

(3) **Crossed Ankle.** (TA) For this position, the sniper crosses his ankles, sits down, and slides his feet forward. Bending at the waist, he places his upper arms inside his knees. As in the other positions, it is mandatory that adjustment of the natural point of aim be accomplished by body movement and not by muscle tension. In the sitting positions, this is done by moving either foot, both feet, or the buttocks until the sights and target are naturally aligned.

d. **Kneeling Position.** (TA) As with the sitting positions, there are three variations of the kneeling position. The low, medium, and high positions. The sniper uses the one best suited to him.

(1) **Assuming Position** (TA)

(a) Kneeling positions require level ground. The sniper kneels on his right knee so that his right leg is parallel to the target. He may then take any one of the three positions. For the low position, the ankle is turned down and the sniper sits on the inside of the ankle. In the medium position, the ankle is straight and the foot stretched out, shoelaces in contact with the ground. The sniper sits on his heel. In the high position, the ankle is straight, toe of the shoe in contact with the ground and curled forward by the body weight. The right side of the buttocks is placed on the right heel making a solid contact. When using these positions, the sniper will be in poor balance if he sits too far to the rear.

(b) From any of these positions, the left leg is extended toward the target with the foot flat on the deck. For maximum support, the toes should be pointed approximately in the direction of the target. To avoid side movement, the left toes are turned slightly to the right by pivoting on the heel. When in position, the left foot may be pushed forward or pulled back slightly to lower or raise the muzzle.

(c) The lower left leg must be positioned properly to provide maximum support for control of the rifle. From a front view, the lower leg should be approximately vertical. In this position, the left leg is using the bone principle to support the weight of the body.

(d) The right elbow is normally held shoulder high to make a pocket for the butt of the rifle. The elbow may be lower if a pocket can be formed without the rifle butt slipping out of the shoulder. The left arm supports the rifle so it is important to notice the relative positions of the various parts of the arm. On the upper arm, there is a flat surface just behind the elbow. This position of the arm is placed against the similarly flat spot on the right side of the left knee. Placing the flat surfaces of the arm and knee together causes the elbow to be forward of the knee, and allows the weight of the body to be transferred forward to the left leg. The left leg must be placed under the rifle to achieve maximum support. There is daylight between the sling and the bend of the left elbow to indicate support of the forearm from the upper arm. The sling supports the bones and, in turn, the bones support the rifle. Approximately 60 percent of the weight of the body is transmitted forward to the left leg, reducing the strain on the right foot and leg and resulting in a relaxed position.

e. Standing Position. The standing position is the least steady and the most difficult to master. However, with proper observation of the fundamentals, excellent results can be obtained.

(1) Assuming Position. The sniper faces his target, executes a right face, and spreads his feet comfortably. His feet must be level to obtain a natural point of aim. With his right hand at the small of the stock, he places the rifle butt high against his shoulder so that the eyepiece of the telescopic sight is level with his eyes. He holds his right elbow high to form a shoulder pocket. This also permits him to exert a strong upward and rearward pressure with his right arm and hand. He holds most of the rifle weight with his right arm and places his left hand under the rifle in a position to best support and steady the rifle. He distributes his weight evenly to both hips. The stock weld for standing is very seldom the same as for the other positions because of the position of the sniper's head. To maintain proper eye relief, the cheek must be placed against the stock in the same way and at the same place everytime. Each sniper can, by practice, determine his proper spot weld.

(2) Shooting in Wind. In the standing position, since the entire body is exposed, the rifle will move considerably with changes in wind velocity or direction. Under this condition, the sniper must wait for a lull or for a period of constancy in wind velocity and direction. While waiting, he allows his body to move with the wind, but at the lull, he quickly acquires the correct sight picture, executes a controlled trigger pressure, and fires a well-aimed shot.

(3) Holding Exercises. Holding exercises will greatly improve the sniper's proficiency from the standing position. In these exercises, the sniper dry fires and remains in position for a specified period, gradually increasing the time from 30 seconds to 1 minute as training progresses. To avoid excessive fatigue, no more than 20 repetitions should be conducted during one training session.

3. Supported Positions. During training in fundamentals, positions are taught in a step-by-step process by which the sniper is guided through a series of precise movements until he obtains the correct position. The purpose of this is to ensure that he knows and correctly applies all of the factors which can assist him in holding the rifle steady. He will gradually become accustomed to the feel of the positions through practice, and will know instinctively whether his position is correct. This is particularly important in combat, since the sniper must be able to assume positions rapidly. There are any number of intermediate positions a sniper in combat might use before assuming his final position. Consequently, he must know that his position is correct without having had to follow a set sequence of movement. In combat situations, the sniper must be prepared to enhance the stability of the position he chooses by adapting it to any available artificial support.

a. Fundamentals Applicable to All Supported Positions. On the battle-field, a sniper must assume the steadiest position which still provides observation of the target area and cover or concealment. Considering the many variables of terrain, vegetation, and tactical situations, there are innumerable positions that might be used. However, in most instances, field supported sniper positions will be variations of those learned during basic marksmanship training and annual known-distance range firing. In assuming a field supported position, the sniper should:

(1) Use any available support; empty sandbags (carried and filled when time permits), rocks, logs, pack, dikes, tree branches, or a sharp rise in terrain are examples.

(2) Avoid touching any part of the support with the rifle barrel.

(3) Adjust position to fit the support.

(4) Use prone position whenever possible.

b. Stability of Weapon. In supported positions, factors affecting weapon stability are identical to those taught during marksmanship training on unsupported positions. The stability of the rifle and the effectiveness of the sniper will be increased by four holding factors.

(1) Grip of the Right Hand. The right hand is placed on the small of the stock. The right thumb is curved over the stock and the forefinger is placed on the trigger without touching the side of the stock. The remaining fingers of the right hand are curled around the small of the stock.

(2) Position of the Cheek (Stock Weld). A conscious effort must be made to place the cheek at the same spot on the stock each time the weapon is fired.

(3) **Placement of Elbows.** The placement of the elbows provides balance to the position. Correctly positioned, the right elbow helps form a pocket in the shoulder for the rifle butt. The exact location of the elbows varies with each position. So long as the sniper applies the fundamentals of maximum support for his rifle, he should be permitted to adjust his elbows to his own preference.

(4) **Position of Left Hand.** The location of the left hand, in supported positions, varies with each position and situation. The sniper must adjust the left hand to his preference within scope of holding fundamentals.

4. **Bolt Operation.** Once the sniper has adopted the position, he wishes and fires a shot, he must then reload. To ensure that the sniper does not modify his position, reloading should be done without moving any part of the body except the right forearm and hand. The sniper must practice reloading with the butt kept at the shoulder and the left and right elbows remaining in position as this will ensure that there is no shift in the point of aim. (TA-Instructor goes through each step of bolt operation with the use of appropriate slides). The sniper must also practice cautious reloading. This involves moving the bolt very slowly while canting the rifle slightly to the right ensuring that the spent casing does not fly through the air or remain inside the chamber, but in fact drops down beside the rifle. To improve concealment during reloading, the sniper's face veil can be lifted up over the rifle and telescopic sight when initially getting into the shooting position. The face veil then acts as a screen to cover the movement of the hand, the bolt and the ejected cartridge case.

OPPORTUNITY FOR QUESTIONS AND COMMENTS

SUMMARY

1. **Reemphasize.** During this period of instruction we have discussed the seven factors common to all positions; left hand and elbow, rifle butt in the pocket of the shoulder, grip of the right hand, right elbow, spot weld, breathing and relaxation.

We then covered each of the five shooting positions used by the sniper and the variations of each. We stressed the fundamentals applicable to all supported positions.

In conclusion, we discussed the proper technique used to operate the bolt with the least amount of movement.

2. **Remotivate.** Some of you as snipers will have more difficulty in assuming a particular position than will others. So long as the sniper applies the fundamentals of relaxation and maximum support for his rifle, he as accomplished half of his mission.

SI 0003
L, D, A

_____
DATE

SIGHT ADJUSTMENT, TELESCOPIC SIGHT, REDFIELD
3X9 VARIABLE, WITH ACCU-RANGE

DETAILED OUTLINE

INTRODUCTION   (30 Min.)

1. <u>Gain Attention</u>. The sniper will accomplish nothing if he fires a round from the most accurate sniper weapon in the world and that rifle has an improper sight adjustment on the telescopic sight.

2. <u>Motivate</u>. The M40A1 sniper rifle coupled with the proper sight adjustment on the telescopic sight is capable of making a first round kill at 1,000 yards.

3. <u>State Purpose and Main Ideas</u>.
    a. <u>Purpose</u>. The purpose of this period of instruction is to provide the student with the knowledge of the proper means of adjusting the telescopic sight, rifle, 3x9 variable with accu-range.
    b. <u>Main Ideas</u>. The main ideas to be discussed are the following:
        (1) Description of Telescopic Sight Adjustments
        (2) Elevation and Windage Rule
        (3) Scale Reading
        (4) Marking the Sights
        (5) Sight Changes

4. <u>Learning Objectives</u>. At the conclusion of this period of instruction, the student without the aid or references, will be able to:
    a. Proficiently apply the elevation and windage rule.
    b. Proficiently use the elevation and windage scales.
    c. Make proper and accurate sight adjustments.

<u>TRANSITION</u>. Prior to zeroing his rifle, the sniper must understand how to adjust for focus and to mount the telescopic sight. In addition, he must understand the use of the elevation and windage rule, how to use the elevation and windage scales and how to make sight adjustments.

<u>BODY</u>.

1. <u>Description of telescopic sight adjustments</u>.
    a. <u>General</u>.
        (1) The telescopic sight has an elevation and windage turret assembly for making sight adjustments, both of which are identical in appearance and movement. The windage and elevation turret assembly both have a dial with an arrow which indicates the direction of movement that is to be followed when moving the strike of the bullet up or down, left or right. There are no distinct clicks noticeable when adjustments are made, making it necessary to set the sights by eye.

2. <u>Elevation and Windage Rule</u>.
    a. The elevation and windage adjustments are graduated in ½ minute of angle. Therefore, a ½ minute adjustment in elevation or windage will move the strike of the bullet ½ inch on the target for each 100 yards of range.

3. <u>Scale Reading. (TA#1)</u>.
    a. The scale on each outer disc has 33 index lines each of which represent a ½ minute of movement, while the windage and/or elevation screw has one index line. One revolution of the adjusting screw equals thirty-three ½ minutes (thirty-three ½ inches) or 16½ full minutes (16½ inches). It is imperative, when recording zeroes, to read the scale by counting the number of index lines on the dial from the fully down position.

4. <u>Marking the Sights</u>.
    a. Sight settings that fall within the first complete rotation (from fully down) of the elevation screw, should be indexed in some manner. Likewise, the elevation zeroes for 200-, 300-, 500- and 600 yards should be marked to expedite sight settings in the field. Since the sniper carries his rifle with a 600 yard zero set on it, he can make rapid adjustments for shorter ranges, and by using the same lines as reference points, he can also adjust for ranges to 1000 yards. The windage turret should be marked to indicate his true 600 yard zero.

5. <u>Sight Changes</u>.
    a. To make sight changes, the sniper first determines the distance from the center of his shot group to the desired location. The distance in elevation is determined vertically, while distance in windage is determined horizontally. These distances are converted to minutes (inches) by using the elevations and windage rules. To raise the strike of the bullet, the elevation adjusting screw is rotated <u>counterclockwise</u>. To move the strike of the bullet right, the adjusting screw is rotated <u>counterclockwise</u>. Conversely, to move the strike of the bullet down, and/or left, the respective adjusting screw is rotated <u>clockwise</u>.

<u>OPPORTUNITY FOR QUESTIONS</u>

<u>SUMMARY</u>

1. <u>Reemphasize</u>. During this period of instruction, we have discussed the description of the telescopic sight adjustment, both in windage and elevation. Initially, we discussed the arrow on both the windage and elevation turret assemblies, which indicates the direction of movement that is to be followed when moving the strike of the bullet up or down, left or right. We stressed the importance of setting the sights by eye, since there were no distinct audible clicks when sight adjustments are made.
    We discussed the elevation and windage rule, learning that all adjustments were graduated in ½ minute of angle allowing the sniper to move the strike of the bullet ½ inch for each 100 yards of range he is, away from the target.
    In the section on scale reading, we covered that there were 33 index lines on each outer disc which represented a ½ minute of movement.
    We discussed the importance of marking the sights for elevation zeroes at 200-, 300-, 500-, and 600 yards as well as the windage zero at 600 yards only.

In conclusion, sight changes were discussed. To make accurate sight changes, the sniper must determine the distance from the center of his shot group to the desired location, and then make appropriate sight changes in elevation and windage to place the next round in the center of the target.

2. Remotivate. Proficiency in making bold, but accurate sight adjustments will be the deciding factor, determining whether you, as the sniper, will achieve that first round kill.

UNITED STATES MARINE CORPS
SCOUT/SNIPER INSTRUCTOR SCHOOL
Marksmanship Training Unit, Weapons Training Battalion
Marine Corps Development and Education Command
Quantico, Virginia 22134

SI 0004
L, D, A
_____
DATE

ZEROING THE TELESCOPIC SIGHT 3X9
<u>VARIABLE WITH ACCU-RANGE, REDFIELD</u>

<u>DETAILED OUTLINE</u>

<u>INTRODUCTION</u>   (60 Min.)

1. <u>Gain Attention</u>. An individual who meets all of the requirements to
qualify as a Marine Sniper, but is unable to apply what he is taught, will
never be a qualified sniper.

2. <u>Motivate</u>. The sniper's ability to become proficient in zeroing his rifle
will determine whether or not he makes that first round kill at ranges out to
 1,000 yards.

3, <u>State Purpose</u>.
     a. <u>Purpose</u>. The purpose of this period of instruction is to provide the
student with the knowledge of becoming proficient in zeroing the telescopic
sight, 3x9 variable with accu-range.
     b. <u>Main Ideas</u>. The main ideas to be discussed are the following:
          (1) Principles of Zeroing
          (2) Methods of Zeroing
          (3) Field Expedient Zeroing

4. <u>Learning Objectives</u>. At the conclusion of this period of instruction, the
student, without the aid of references, will be able to:
     a. Relate a clear concise understanding of bullet path, point of aim and
trajectory.
     b. Proficiently apply the methods of zeroing.
     c. Apply the two methods of field expedient zeroing.

<u>TRANSITION</u>. To understand the principles of zeroing, the sniper should have
a basic knowledge of the relationship between the path of the bullet in flight
and the line of aim.

1. <u>BODY</u>. <u>Principles of Zeroing</u>.
     a. <u>Bullet Path and Point of Aim</u>. In flight, a bullet does not follow a
straight line but travels in a curve or arc which is called trajectory. (TA#1)
The maximum height of a bullet's trajectory depends on the range to the target.

The greater the distance a bullet travels before impact, the higher it must travel in it's trajectory. On the other hand, the line of aim is a straight line from the eye through the telescopic sight to the aiming point or target. After the bullet leaves the rifle, it is initially moving in an upward path, intersecting the line of aim; as the bullet travels, it begins to drop and eventually intersects the line of aim again. The range at which this intersection occurs is the zero for that range setting.

    b. <u>Definition of Zero</u>. The zero of a rifle is the sight setting in both elevation and windage required to place a shot, in the center of a target at a given range when no wind is blowing.

2. <u>Methods of Zeroing</u>. The zeroing procedure to be discussed will be used to establish zeroes for various desired ranges. It is imperative that all zeroes be entered in the snipers log book. When operating in areas where climatic conditions will have an adverse affect on the weapons performance, the weapon and telescopic sight should be returned for appropriate maintenance every 2 weeks to check the zero, rezero, and perform other required maintenance.

    a. <u>Boresighting</u>. The initial zeroing phase should start at the 200 yard line. To facilitate getting the first round "on target" quickly, it is imperative that the rifle be boresighted prior to firing. This is done to place the axis of the rifle bore and the telescopic sight on the same focal plane. The correct procedure to boresight a bolt operated rifle is:

      (1) With bolt removed from the rifle, place the rifle on a solid support such as a sandbag.

      (2) Looking through the barrel, adjust the rifle until the desired aiming point is visible through the center of the bore.

      (3) Without disturbing the lay of the rifle, look through the telescope and observe the position of the crosshairs in relation to the aiming point.

      (4) If the crosshairs do not coincide with the aiming point as seen through the bore, adjust the elevation and windage screws/knobs until the crosshairs quarter the aiming point.

      (5) Resight through the bore to ensure the rifle has not moved.

      (6) Raise the elevation on the sight 3½ minutes. The rifle should now be fired to confirm the rifle and telescopic sight alignment.

<u>TRANSITION</u>. After the weapon is boresighted, depending on the range to the target, an adjustment must be made in elevation.

    b. <u>Elevation Adjustments with Respect to Boresight</u>. This adjustment in elevation is necessary to compensate for the drop of the bullet for that particular range. Adjustments for the following ranges are:
    (TA #2)

<u>TRANSITION</u>. After the sniper has boresighted his rifle, he now fires triangulation to obtain his true zero.

    c. <u>Zeroing by Triangulation</u>. This is done by assuming the desired position and firing three 3-round shot groups. Sight corrections are made after each three rounds (if required) to move the shot groups into the center of the target. After the 200-yard zero is obtained, the sniper whould use the same procedure to arrive at a zero for 300 and every 100 yards of range out to 1,000 yards.

<u>TRANSITION</u>. As the range increases, the sniper must increase the elevation on the sight.

d. <u>Normal Come-Up in Elevation</u>. Each sniper will determine his normal come-up as he zeroes his weapon for each succeeding range. The average come-up in elevation between ranges is as follows:
(TA #3)

<u>TRANSITION</u>. In many instances, known-distance ranges will be unavailable to the sniper for zeroing or rezeroing his weapon. In such cases, field expedient methods will be used to determine correct sight settings. Two methods in general use are the 900-inch and the unknown-distance method.

3. <u>Field Expedient Zeroing</u>.
    a. <u>900-Inch Method</u>. (TA #4) This sample target is designed for establishing 300-, 500-, and 750-yard zeroes while firing at 900 inches (25 yards). Targets may be constructed from available material with black bull's eyes painted to use as aiming points and a plain circle is drawn a prescribed distance above each to act as impact areas.
        (1) To establish or verify the zero for any of the ranges shown in the illustration, you must:
            (a) Quarter the black spot with the crosshairs of the telescopic sight.
            (b) Fire a round and observe the impact of the bullet.
            (c) Make the necessary adjustments on the telescopic sight to bring the impact of the bullet into the upper circle for the particular range.
            (d) Fire again and observe. Continue adjustments until several shots strike within the upper circle.
        (2) The rifle is now zeroed for that range. The sniper may then apply normal come-up or come-down to establish zeroes at other ranges.
        (3) Field expedient targets may also be produced for use with various rifle and telescopic sight combinations. It is necessary in each case to arrive at the correct distance between the lower and upper bull's-eye. To accomplish this, it is necessary to:
            (a) Use a rifle telescopic sight combination for which a zero has been previously established on a known-distance range.
            (b) Using that established sight setting for that range, fire a three-shot group at 900 inches ensuring that the crosshairs are quartered on the lower bull's-eye.
            (c) Measure the point of impact from the point of aim for each shot fired.
            (d) Add all the measurements and divide the total by the number of shots fired to establish an average.
            (e) The resulting average in inches will then be used to establish the center point for the upper bull's-eye.
    b. <u>Unknown-Distance Zeroing</u>. When a known-distance range is not available the zero of a telescopic sight can be checked/established by the following procedure:
        (1) Pick out an aiming point in the center of an area where the observer can see the strike of the bullet. This can be a hillside, a brick house or stone house, or any dry surface where the strike of the bullet can be observed. The range to this target must be determined by map, from the range card of another weapon, or by measurement.
        (2) Boresight the rifle and fire one shot at the aiming point.
        (3) An observer notes the bullet strike and gives the elevation and windage change necessary to bring it to the point of aim. He does this by estimating the distance of the bullet strike right or left, high or low, and

converts these distances to minutes of change by dividing the error in inches by the number of inches 1 minute will move the strike at the given range.

OPPORTUNITY FOR QUESTIONS

SUMMARY

1. <u>Reemphasize</u>. During this period of instruction, we have discussed the principles of zeroing and the relationship between bullet path, point of aim and trajectory.

We defined zero as being that sight setting in both elevation and windage required to place a shot, or the center of a shot group, in the center of a target at a given range when no wind is blowing.

In the section on methods of zeroing, it was mentioned that all zeroes must be entered in the sniper's log book and that the weapon needed the proper maintenance performed approximately every two weeks when operating in adverse climates and rezeroed.

We discussed step-by-step procedures for boresighting and the elevation adjustments involved, zeroing by triangulation, normal come-ups in elevation, field expedient zeroing and unknown-distance zeroing.

2. <u>Remotivate</u>. The sniper weapon in the hands of a qualified Marine, proficient and knowledgeable in the principles and methods of zeroing and their application, is capable of delivering a devastating blow to the ranks of the enemy.

"COME UPS" FOR LAKE CITY AMMUNITION
M40A1 SNIPER RIFLE

YARDS	COME UP
100-200	1-1/2 min
200-300	3
300-400	3-1/2
400-500	4
500-600	4-1/2
600-700	5
700-800	5
800-900	6
900-1000	7

SI0005
L, D, A

_____

(DATE)

CAMOUFLAGE

DETAILED OUTLINE

INTRODUCTION

1. **Gain Attention.** Most uninformed people envision a sniper to be a person with a high powered rifle who either takes pot shots at people from high buildings or ties himself in a coconut tree until he's shot out of it. But to the enemy, who knows the real capabilities of a sniper, he is a very feared ghostly phantom who is never seen, and never heard until his well aimed round cracks through their formation and explodes the head of their platoon commander or radio man. A well trained sniper can greatly decrease the movement and capabilities of the most disciplined troops because of the fear of this unseen death.

2. **Motivate.** This gives one example of how effective snipers can be and how the possibility of their presence can work on the human mind. But, marksmanship is only part of the job. If the sniper is to be a phantom to the enemy, he must know and apply the proper techniques of camouflage. He cannot be just good at camouflage. He has to be perfect if he is to come back alive.

3. **State Purpose and Main Ideas**

   a. **Purpose.** The purpose of this period of instruction is to provide the student with the basic knowledge to be able to apply practically, the proper techniques of camouflage and concealment needed to remain undetected in a combat environment.

   b. **Main Ideas.** The main ideas which will be discussed are the following:

      (1) Target Indicators
      (2) Types of Camouflage
      (3) Geographical Areas
      (4) Camouflage During Movement
      (5) Tracks and Tracking

4. **Learning Objectives.** Upon completion of this period of instruction, the student will:

   a. Camouflage his uniform and himself by using traditional or expedient methods as to resemble closely the terrain through which he will move.

b. Camouflage all of his equipment to present the least chance of detection.

c. Know and understand the principles which would reveal him in combat and how to overcome them.

d. Understand the basic principles of tracking and what information can be learned from tracking.

BODY

1. Target Indicators

a. General. A target indicator is anything a sniper does or fails to do that will reveal his position to an enemy. A sniper has to know and understand these indicators and their prindiples if he is to keep from being located and also that he may be able to locate the enemy. Additionally, he must be able to read the terrain to use the most effective areas of conceal- ment for movement and firing positions. Furthermore, a sniper adapts his dress to meet the types of terrain he might move through.

b. Sound. Sound can be made by movement, equipment rattling, or talking. The enemy may dismiss small noises as natural, but when someone speaks he knows for certain someone is near. Silencing gear should be done before a mission so that it makes no sound while running or walking. Moving quietly is done by slow, smooth, deliberate movements, being conscious of where you place your feet and how you push aside bush to move through it.

c. Movement. Movement in itself is an indicator. The human eye is attracted to movement. A stationary target may be impossible to locate, a slowly moving target may go undetected, but a quick or jerky movement will be seen quickly. Again, slow, deliberate movements are needed.

d. Improper Camouflage. The largest number of targets will usually be detected by improper camouflage. They are divided into three groups.

(1) Shine. Shine comes from reflective objects exposed and not toned down, such as belt buckles, watches, or glasses. The lenses of optical gear will reflect light. This can be stopped by putting a paper shade taped to the end of the scope or binoculars. Any object that reflects light should be camouflaged.

(2) Outline. The outline of items such as the body, head, rifle or other equip- ment must be broken up. Such outlines can be seen from great distances. There- fore, they must be broken up into features unrecognizable, or unnoticable from the rest of the background.

(3) Contrast with the Background. When using a position for concealment, a background should be chosen that will absorb the appearance of the sniper and his gear. Contrast means standing out against the background, such as a man in a dark uniform standing on a hilltop against the sky. A difference of color or shape from the background will usually be spotted. A sniper must therefore use the coloring of his background and stay in shadows as much as possible.

2. Types of Camouflage

   a. **Stick Camouflage.** In using the "grease paint", all the exposed skin should be covered, to include the hands, back of the neck, ears, and face. The parts of the face that naturally form shadows should be lightened. The predominate features that shine, should be darkened, such as the forehead, cheeks, nose and chin. The pattern and coloring that should be used is one that will blend with the natural vegetation and shadows. For jungle or woodland, dark and light green are good. White and gray should be used for snow areas, and light brown and sand coloring for deserts.

      (1) **Types of Patterns.** The types of facial patterns can vary from irregular stripes across the face to bold splotching. The best pattern, perhaps, is a combination of both stripes and splotches What one does not want is a wild type design and coloring that stands out from the background.

   b. **Camouflage Clothing.**

      (1) **The Ghillie Suit.** The ghillie suit is an outstanding form of personal camouflage. It is used by both the British and Canadian Snipers to enable them to stalk close to their targets undetected. The ghillie suit is a camouflage uniform or outer smock that is covered with irregular patterns of garnish, of blending color, attached to it. It also has a small mesh netting sewn to the back of the neck and shoulders, and then draped over the head for a veil. The veil is used while in position to break up the outline of the head, hide the rifle scope, and allow movement of the hands without fear of detection. The veil when draped over the head should come down to the stomach or belt and have camouflaged garnish tied in it to break up the outline of the head and the solid features of the net. When the sniper is walking, he pushes the veil back on his head and neck so that he will have nothing obstructing his vision or hindering his movements. The veil is, however, worn down while crawling into position or while near the enemy. The ghillie suit, though good, does not make one invisible. A sniper must still take advantage of natural camouflage and concealment. Also wearing this suit, a sniper would contrast with regular troops, so it would only be worn when the sniper is operating on his own.

      (2) **Field Expedients.** If the desired components for the construction of a ghillie suit are not on hand, a make-shift suit can be made by expedient measures. The garnish can be replaced by cloth discarded from socks, blankets, canvas sacks, or any other material that is readily available. The material is then attached to the suit in the same way. What is important is that the texture and outline of the uniform are broken. The cloth or any other equipment can be varied in color by using mud, coffee grounds, charcoal, or dye. Oil or grease should not be used because of their strong smell. Natural foliage helps greatly when attached to the artificial camouflage to blend in with the background. It can be attached to the uniform by elastic bands sewn to the uniform or by the use of large rubber bands cut from inner tubes. Care must be taken that the bands are not tight enough to restrict movement or the flow of blood. Also as foliage grows old, or the terrain changes, it must be changed.

c. Camouflaging Equipment

(1) One of the objects of primary concern for camouflage is the rifle. One has to be careful in camouflaging the rifle that the operation is not interfered with, the sight is clear, and nothing touches the barrel. Camouflage netting can be attached to the stock, scope and sling, then garnish tied in it to break up their distinctive outline.
The M-16 and M-14 can be camouflaged in the same way ensuring that the rifle is fully operational.

(2) Optical Gear such as the M-49 scope and binoculars are camouflaged in the same manner. The M49 and stand is wrapped or draped with netting and then garnish is tied into it. Make sure that the outline is broken up and the colors blend with the terrain. The binoculars are wrapped to break their distinctive form. Since the glass reflects light, a paper hood can be slipped over the objective lens on the scope or binoculars.

(3) Packs and Web Gear. Web gear can be camouflaged by dying, tying garnish to it, or attaching netting with garnish. The pack can be camouflaged by laying a piece of netting over it, tied at the top and bottom. Garnish is then tied into the net to break up the outline.

3. Geographic Areas

a. General. One type of camouflage naturally can not be used in all types of terrain and geographic areas. Before operations in an area, a sniper should study the terrain, vegetation and lay of the land to determine the best possible type of personal camouflage.

(1) Snow. In areas with heavy snow or in wooded areas with trees covered with snow, a full white camouflage suit is worn. With snow on the ground and the trees are not covered, white trousers and green-brown tops are worn. A hood or veil in snow areas is very effective. Firing positions can be made almost totally invisible if made with care. In snow regions, visibility during a bright night is as good as in the day.

(2) Desert. In sandy and desert areas, texture camouflage is normally not necessary, but full use of the terrain must be made to remain unnoticed. The hands and face should be blended into a solid tone using the proper camouflage stick, and a hood should be worn

(3) Jungle. In jungle areas, foliage, artificial camouflage, and camouflage stick are applied in a contrasting pattern with the texture relative to the terrain. The vegetation is usually very thick so more dependence can be made on using the natural foliage for concealment.

4. Camouflage During Movement.

a. Camouflage Consciousness. The sniper must be camouflage conscious from the time he departs on a mission until the time he returns. He must constantly observe the terrain and vegetation change. He should utilize shadows caused by vegetation, terrain features, and cultural features to remain undetected. He must master the techniques of hiding, blending, and deceiving.

(1) **Hiding**. Hiding is completely doncealing yourself against observation by laying youself in very thick vegetation, under leaves, or however else is necessary to keep from being seen.

(2) **Blending**. Blending is what is used to the greatest extent in camouflage, since it is not always possible to completely camouflage in such a way as to be indistinguishable from the surrounding area. A sniper must remember tnat his camouflage need be so near perfect that he should fail to be recognized through optical gear as well as with the human eye. He must be able to be looked at directly and not be seen. This takes much practice and experience.

(3) **Deceiving**. In deceiving, the enemy is tricked into false conclusion regarding the sniper's location, intentions, or movement. By planting objects such as ammo cans, food cartons, or something to intrique the enemy, that he may be decoyed into the open where he can be brought under fire. Mannequins can be used to lure the enemy sniper into firing, thereby revealing his position.

5. **Tracks and Tracking**.

a. **General**. Once a sniper has learned camouflage and concealment to perfection, he must go one step further. This is the aspect of him leaving no trace of his presence, activities or passage in or through an area. This is an art in itself, and is closely related to tracking, which can tell you in detail about the enemy around you.

(1) **Enemy Trackers or Scouts**. It is said that the greatest danger to a sniper is not the regular enemy soldier, but in fact, is hidden booby traps, and the enemy scout who can hunt the sniper on his own terms. If an enemy patrol comes across unfamiliar tracks in it's area of operation, it may be possible for them to obtain local trackers. If it is a man's liveli-hood to live by hunting, he will usually be very adept at tracking. What a professional can read from a trail is truly phenominal. Depending on the terrain, he will be able to determine the exact age of the trail, the number of persons in the party, if they are carrying heavy loads, how well trained they are by how well they move, their nationality, by their habits and boot soles, how fast they are moving and approximately where they are at the moment. If a tracker determines a fresh trail to be a party of four, who, but recon and sniper teams move in such small groups behind enemy lines? The enemy will go to almost any extreme to capture or kill them.

(2) **Hiding Personal Signs**. A modern professional tracker who makes his living of trailing lost children, hunters, or escaped convicts. was once asked, "Can a man hide his own trail well enough that a tracker cannot follow him?" The professional tracker answered, "NO, there is no way to hide a trail from a true tracker." The chances in combat of being pitted against a "real tracker" are rare, but all it takes is one time. This is to emphasize the importance of leaving no signs at all for the enemy scout to read. This is done by paying particular attention to where and how you walk, being sure not to walk in loose dirt or mud if it can be avoided, and not scuffing the feet. Walking on leaves, grass, rocks, etc. can help hide tracks. Trails are also made by broken vegetation such as weeds, limbs, scrape marks on bushes, and limbs that have been bent in a certain direction. When moving through thick brush, gently move the brush forward, slip through it, then set it back to it's normal position.

Mud or dirt particles left on rocks or exposed tree roots are a sign of one's presence. Even broken spider webs up to the level of a man's height show movement. In the process of hiding his trail, a sniper must remember to leave no debris such as paper, C-Ration cans, spilled food, etc. behind him. Empty C-Ration cans, can either be carried out, or smashed, buried, and camouflaged. Along this same line, a hole should be dug for excrement, then camouflaged. The smell of urine on grass or bushes lasts for many days in a hot humid environment, so a hole should be dug for this also. One last object of importance, the fired casings from the sniper rifle must <u>always</u> be brought back, for they are a sure sign of a sniper's presence.

(3) <u>Reading Tracks and Signs</u>. To be proficient at tracking takes many years of experience, but a knowledgable sniper can gain much information from signs left by the enemy. For instance, he can tell roughly the amount of enemy movement through a given area, what size units they move in, and what areas they frequent the most. If an area is found where the enemy slept, it may be possible to determine the size of the unit, how well disciplined they are, by the security that was kept, and their overall formation. It can be fairly certain that the enemy is well fed if pieces of discarded food or C-Ration cans with uneaten food in them are found. The opposite will also be true for an enemy with little food. Imprints in the dirt or grass can reveal the presence of crew-served weapons, such as machine guns or morters. Also, prints of ammo cans, supplies, radio gear may possibly be seen. The enemies habits may come to light by studying tracks so that he may be engaged at a specific time and place.

<u>OPPORTUNITY FOR QUESTIONS AND COMMENTS</u>

<u>SUMMARY</u>

1. <u>Reemphasize</u>. During this period of instruction, we have discussed target indicators, types of camouflage, geographical areas, and tracks. Initially, we covered what a target indicator is, and that sound, movement, and improper camouflage make up indicators. Care must be taken that shine, outline, and contrast with the backgouund are eliminated.
We learned the different types of camouflage, such as grease paint, and how to tone down the skin with it, the ghillie suit, used as camouflaged clothing, and field ixpedient measures for camouflaging alothing and equipment. In the section on geographical areas, we learned the different types of camouflage used in the various climate regions.

The fourth area covered was concealment during movement and how to use terrain features. We learned the difference in hiding, blending, and deceiving, and how to use each. We learned of the danger of enemy scouts or trackers, and the importance of leaving no indication of one's presence while on a mission. Lastly, we covered what a sniper can learn from enemy tracks if he is observant enough to see them and takes the time to learn their meaning.

2. <u>Remotivate</u>. The job of a sniper is not for a person who just wants the prestige of being called a sniper. It is a very dangerous position even if the sniper is well trained and highly motivated. Expertise at camouflage to remain unnoticed takes painstaking care, and thoroughness which the wrong type of individual would not take time to do. If you are to be successful at camouflage and concealment, it takes a double portion in carefulness on your part, if you are to come back alive.

UNITED STATES MARINE CORPS
SCOUT/SNIPER INSTRUCTOR SCHOOL
Marksmanship Training Unit, Weapons Training Battalion
Marine Corps Development and Education Command
Quantico, Virginia 22134

SIO006
L, D

_____
(DATE)

ELEVATION AND WINDAGE HOLDS

DETAILED OUTLINE

INTRODUCTION

1. Gain Attention. You've been sitting in position for three hours maintaining observation over your assigned area. You have a 600 yd zero on your rifle. All of a sudden, a target presents itself, an enemy soldier has just stepped out of a door way to look around quickly. You estimate the range to be 400 yds, but you do not have time to adjust your scope, so you aim for the center of the chest and fire. Your partner tells you that you hit just over your targets head, and he escaped.

2. Motivate. Today you are going to learn about elevation and windage holds. These are the tools which, if you had understood them, would have helped you to send that enemy soldier to the great battlefield in the sky.

3. State Purpose and Main Ideas.

    a. Purpose. To give the student an understanding of how to hit a target at a range other than six hundred yards, using a six hundred yard zero and how to compensate for the strength of the wind without ajusting the windage knob on his scope.

    b. Main Ideas. Explain through a discussion of the trajectory of a round, and the value of the wind, where a round will hit using any given aiming point and a preset zero on the rifle.

4. Learning Objectives. Upon completion of this period of instruction, the student will be able to:

    a. Understand and demonstrate proper elevation holds for any range from 100 yards to 800 yards with a 600 yard zero.

    b. Understand and demonstrate proper windage holds for any range from 100 yards to 1000 yards.

<u>TRANSITION</u>. In recruit training or at the Basic School, while undergoing marksmanship training, you all probably heard horror stories about the use of "Kentucky" windage, or the art of "shooting while they aint". Let's now look at the system of shooting where they aint to hit them where they <u>are</u>.

<u>BODY</u>.

1. <u>Hold-Off</u>.

    a. The <u>holdoff</u> is the hold for elevation and is defined as the procedure used to hit a target at ranges other than the range for which the rifle is zeroed. When firing at a range greater than that for which the rifle is zeroed, the bullet will hit below the point of aim. At lesser ranges, the bullet will hit higher than the point of aim. Understanding this, and with a knowledge of trajectory and bullet drop, the sniper will be able to hit a target at ranges other than that for which the rifle was zeroed.

    b. Place TP #1 on overhead projector. Trajectory and Bullet Drop.

    c. The transparency illustrates trajectory for various ranges and the bullet drop for 100 yds past the ranges. For combat firing, the sniper normally keeps a 600 yard zero on his rifle. With this setting, he can engage targets up to 800 yards by holding over the target. At ranges under 600 yds, he holds under the target.

    d. Place TP #2 on the overhead projector. Correct Holds For Various Ranges With Sights Set For 600 Yards.

2. <u>Holding for Wind</u>.

    a. Place TP on over-head projector, Wind Conversion Table.

    b. The sniper may holdoff to compensate for the effect of the wind. This transparency illustrates the effect of the wind on the bullet in inches. It also indicates the necessary sight setting required to compensate for these wind effects, providing the sniper has time to make such adjustments.

    c. The sniper holds off for the wind by aiming into it, if the wind is from the right, he aims to the right of his target the required distance, if the wind is from the left, he aims to the left. Adjustments for wind are always based on the estimate of it's velocity, and constant practice is the only way for the sniper to develop proficieny in wind reading.

    d. If a miss is fired, and the impact of the round observed, the sniper will note the lateral distance of his error and refire, holding off that distance in the opposite direction.

<u>OPPORTUNITY FOR QUESTIONS AND COMMENTS</u>   (1 Min)

SUMMARY

1. __Reemphasize__. We have just examined the two methods for hitting a target at ranges other than that for which the rifle is zeroed and in varying wind conditions, when time does not allow adjusting the sights.

2. __Remotivate__. Remember, the difference between a hit or a miss in a fast moving situation may well depend on your ability to execute a holdoff.

UNITED STATES MARINE CORPS
SCOUT/SNIPER INSTRUCTOR SCHOOL
Marksmanship Training Unit, Weapons Training Battalion
Marine Corps Development and Education Command
Quantico, Virginia 22134

SIO007
L, D

_____
(DATE)

SIGHTING, AIMING AND TRIGGER CONTROL

DETAILED OUTLINE

INTRODUCTION

1. **Gain Attention.** The first of the basic marksmanship fundamentals taught to the shooter are sighting, aiming and trigger control. The reason for this is that without an understanding of the fundamentals, a sniper will not be able to accomplish his primary mission.

2. **Motivate.** During your six weeks here at the school, you will be taught individual movement, cover and concealment, map reading and many other related sniper skills, in addition to marksmanship. By the time you graduate, you will be able to move to a sniper position, fire a shot and withdraw, all without being observed. However, all this will be to no avail if, because you do not understand the principles of sighting, aiming and trigger control, you miss your target when you do fire.

3. **State Purpose and Main Ideas.**

    A. **Purpose.** To introduce the student to the principles of sighting, aiming and trigger control as applied to the M40A1 sniper rifle with telescopic sight.

    B. **Main Ideas.**

        (1) Sighting and aiming will be discussed in three phases:

            (a) The relationship between the eye and the sights

            (b) Sight picture

            (c) Breathing

        (2) Trigger control will be explained through smooth action, interupted pull, concentration and follow-through.

4. **Learning Objectives.** Upon completion of this period of instruction, the student will be able to:

    A. Understand and demonstrate the sighting error known as "shadow effect";

    B. Understand and demonstate "Quartering" the target.

    C. Understand and demonstrate proper trigger control.

TRANSITION. The arrangement of an optical sight allows for aiming without recourse to organic rifle sights. The role of the front sight in a telescope is fulfilled by the crosshairs. Because the crosshairs and the image of the target are in the focal plane of the lens, the shooter can use both at the same time and with equal clarity.

BODY

A. Aiming

(1) Relationship between the eye and the sights. In order to see what is required during aiming, the shooter must know how to use his eye. Variations in the positions of the eye to the telescope will cause variations in the image received by the eye. The placement of eye in this respect is called eye relief. Proper eye relief is approximately 2-3 inches from the exit pupil of the telescope, and can be determined to be correct when the shooter has a full field of view in the telescope with no shadows. If the sniper's eye is located without proper eye relief, a circular shadow will occur in the field of vision, reducing the field size, hindering observation, and, in general, making aiming difficult. If the eye is shifted to one side or another of the exit pupil, cresent shaped shadows will appear on the edges of the eyepiece (TP #1 Shadow Effects). If these cresent shaped shadows appear, the bullet will strike to the side away from the shadow. Therefore, when the sniper has a full field of view and is focusing on the intersection of the crosshairs, he has aligned his sight.

(2) Sight Picture. With the telescopic sight, this is achieved when the crosshairs are centered on the target, and the target has been quartered. (Place on TP #3, Sample Sight Pictures). This transparency shows samples of different types of sight pictures. In each case, you can see that the target has been quartered to maximize the chance of a first round hit.

(3) Breathing. The control of breathing is critical to the aiming process. If the sniper breathes while aiming, the rising and falling of his chest will cause the muzzle to move vertically. To breath properly during aiming, the sniper inhales, then exhales normally and stops at the moment of natural respiratory pause. The pause can be extended to 8-10 seconds, but it should never be extended until it feels uncomfortable. As the body begins to need air, the muscles will start a slight involuntary movement, and the eyes will loose their ability to focus critically. If the sniper has been holding his breath for more than 8-10 seconds, he should resume normal breathing and then start the aiming process over again.

B. Trigger Control

(1) The art of firing your rifle without disturbing the perfected aim is the most important fundamental of marksmanship. Not hitting where you aim is usually caused by the aim being disturbed just before or as the bullet leaves the barrel. This can be caused by jerking the trigger or flinching as the rifle fires. A shooter can correct these errors by using the correct technique of trigger control.

(2) Controlling the trigger is a mental process, while pulling the trigger is a physical one. Two methods of trigger control used with the sniper rifle are smooth action and interrupted pull.

(a) Using the smooth action method, the shooter takes up the initial pressure, or free play, in the trigger. Then, when the aim is perfected, increases the pressure smoothly until the rifle fires.

(b) When using the interrupted method of trigger pull, the shooter takes up the initial pressure and begins to squeeze off the shot when the aim is perfected. However, because of target movement or weapon movement, he pauses in his trigger squeeze until the movement stops, then continues to squeeze until the weapon fires.

(c) Trigger Control Developed as a Reflex. The shooter can develop his trigger control to the point that pulling the trigger no longer requires conscious effort. The shooter will be aware of the pull, but he will not consciously be directing it. A close analogy to trigger control can be found in typing. When first learning to type, the typist reads the alphabetic letter to be typed, mentally selects the corresponding key, and consciously directs a finger to strike the key. After training and practice, however, the typist will see the letter which has to be typed and the finger will hit it automatically. This then, is a conditioned reflex; conditioned because it was built in and reflex because it was not consciously directed. The same type of conditioned reflex can be developed by the sniper. When he first starts firing, he must consciously direct his finger to squeeze the trigger as soon as the aim is perfected. As a result of training, however, a circuit will be established between the eye and the trigger finger. The eye, seeing the desired sight picture, will cause the finger to squeeze the trigger without conscious mental effort. The shooter, like the typist is aware of pressure against the trigger, but is not planning or consciously directing it.

(d) Developing Trigger Control. In all positions, one of the best methods for developing proper trigger control is through dry firing, for here the shooter is able to detect his own errors without having recoil conceal undesirable movements. Only through patience, hard work, concentration and great self-discipline will the mastery of trigger control be achieved.

(3) Factors Affecting Trigger Control.

(a) Concentration. The shooter's concentration should be focused on the perfection of aim, as trigger control is applied. Concentration defined as the will to demand obedience, is the most important factor in the technique of trigger control.

(b) Placement of the Trigger Finger and Grip on the Rifle. The finger should be placed on the trigger in the same place each time. Only through practice can the shooter determine which part of his finger should go on which spot on the trigger. Any position of the finger in relation to the trigger is acceptable so long as the shooter can pull the trigger straight to the rear.

Moreover, in order to achieve a smooth, consistent trigger squeeze, the stock must be grasped firmly and in the same place each time.

        (c) **Follow-Through**. Follow-through means doing the same things after each shot is fired, thereby insuring that there is no undue movement of the rifle before the bullet leaves the barrel. The shooter continues to hold his breath, to focus on the crosshairs and to practice trigger control even though the rifle has already fired. By doing this, the shooter can detect any errors in sight alignment and sight picture and he can correct them after follow-through has been completed.

<u>OPPORTUNITY FOR QUESTIONS AND COMMENTS</u>   (1 MIN)

<u>SUMMARY</u>   (1 MIN)

1. <u>Reemphasize</u>. We have just covered the marksmanship fundamentals of sighting, aiming and trigger control, and how to apply them properly.

2. <u>Remotivate</u>. Your ability to hit a target at any range and under any conditions will be a measure of how well you have practiced and mastered these principles.

UNITED STATES MARINE CORPS
SCOUT/SNIPER INSTRUCTOR SCHOOL
Marksmanship Training Unit, Weapons Training Battalion
Marine Corps Development and Education Command
Quantico, Virginia 22134   SI0008
                                            L, D, A
EFFECTS OF WEATHER
                                            _____
                                                (DATE)
DETAILED OUTLINE

INTRODUCTION

1. <u>Gain Attention</u>. If a sniper can observe a flag or any clothlike
material similar to a flag hanging from a pole, he may estimate the angle
(in degrees) formed at the juncture of the flag or cloth and it's support
and divide this angle by the constant number "4" to get the wind velocity
in miles per hour. After finding velocity, then multiply range x velocity,
divided by 15. That equals minutes of movement required.

2. <u>Motivate</u>. Refer to attention gainer, point out the advantages of the
two formulas for first shots, since the sniper will get no sighters.

3. <u>Purpose and Main Ideas</u>.

    a. <u>Purpose</u>. The purpose of this period of instruction is to teach
you the effects of weather as it effects the shooter, rifle and ammunition.

    b. <u>Main Ideas</u>. The main ideas which will be discussed are the
following:

        (1) Effects of Wind
        (2) Effects of Light
        (3) Humidity and Temperature

4. <u>Learning Objectives</u>. At the conclusion of this period of instruction, the
student, without the aid of references, will be able to:

    a. Become proficient in estimating wind velocity and applying that to
sight changes.

    b. Explain the effect temperature, humidity and mirage have on the
bullet.

    c. Proficiently judge light and know how that effects you as a sniper.

<u>TRANSITION</u>. Effects of weather can cause serious error in the strike of the
bullet. The wind, light, temperature, and humidity all have some effect on
the bullet, the sniper, or both. Under average conditions, some weather
effects, such as temperature and humidity, are relatively insignificant, but
since sniping is sometimes done under extremes of weather, all effects must
be considered.

BODY

1. __Effects of Wind__. Wind, usually, is the greatest weather problem. Wind effect on the bullet increases as the wind velocity increases, as wind approaches a cross direction, and as the target distance increases. Wind also affects the sniper, particularly in the standing position. The stronger the wind, the more difficulty he has in holding his rifle steady. This effect can be offset by use of supported positions.

    a. __Classification of Winds__. Winds are classified according to the direction from which they are blowing in relation to the direction of fire. The clock system (TA #1) is used to indicate this direction. A wind blowing from right to left or left to right across the sniper's front is called a "3 or 9 o'clock wind" respectively and will have the most effect on the flight of the bullet to the target. You are always firing toward 12 o'clock, with 2 o'clock to your rear, 3 o'clock to your right and 9 o'clock to your left. The direction from which the wind is blowing also denotes the fractional value of the wind in relation to the wind's total velocity. Winds from either flank are "full value" those from the oblique are "half value", and those from the front or rear are "no value" winds. A half value wind affects the bullet approximately half as much as a full value wind. That is, a 1 o'clock wind having a velocity of 10 miles per hour is equivalent to a 5 mile per hour 3 o'clock wind.

    b. __Wind Velocity__. There are three common field expedient methods of approximating wind velocities. Since the tactical situation may preclude the use of some methods, snipers must be thoroughly familiar with all.

        (1) The flag method was explained in the opening portion of the class. (Re-explain) (TA #2)

        (2) __Pointing Method__. TA # 3) A piece of paper or other light material is dropped from the sniper's shoulder. He points to the spot where it lands and estimates the angle between his arm and body. This figure is then divided by the constant number of "4".

        (3) __Observation Method__. If the tactical situation prevents the use of the methods previously mentioned, snipers can use these guides:

        (a) A wind under 3 miles per hour can hardly be felt, but it causes smoke to drift.

        (b) A 3-5 mile per hour wind is felt lightly on the face.

        (c) A 5-8 mile per hour wind keeps tree leaves in constant motion.

        (d) An 8 - 12 mile per hour wind raises dust and loose paper.

        (e) A 12-15 mile per hour wind causes small trees to sway.

(4) <u>Reading a Mirage</u>. "Mirage", as used here, applies to the (TA #4) atmospheric phenomenon in which air appears to move in ascending waves; the sun heating the earth's surfaces causes heat waves in the same manner as heating a piece of metal does. These waves appear to ripple or shimmer and rise straight up from the ground on a calm day. Any movement of air, however, will bend these waves in the direction of the air flow. Mirage is most pronounced on a bright, clear day and becomes more pronounced when viewed over unbroken terrain or where vegetation is sparse or lacking. Mirage always appears to be flowing from either 3 or 9 o'clock, or appears to be 'boiling' and stationary when there is no wind or when the wind is blowing from 6 to 12 o'clock. If the mirage is flowing from the right, the wind is coming from either 1, 2, 3, 4, or 5 o'clock. To determine which, the spotting scope is turned slowly to the right, but not beyond 3 o'clock. As the scope turns, the mirage may 'boil'. If so, the direction in which the scope points is the direction 2 or 3 o'clock. If no 'boil' occurs, turn the scope left, but not beyond 9 o'clock. When the boil is seen, the direction in which the rear of the eyepiece of the scope is pointing is the direction from which the wind is blowing, 4 or 5 o'clock.

(a) <u>Determination of Windage Adjustment</u>. After finding the wind direction and velocity, the windage correction bo be placed on the sight is determined by the formula: RxV/15=number of minutes of windage to be placed on the sight for a full value wind. R = range in hundreds of yards, V = velocity in miles per hour, and the number 15 is constant. For half value winds, divide the answer by 2. The formula is accurate up to ranges of 500 yards, but beyond 500 yards, it varies due to velocity loss. However, by changing the constant 15, the correct windage may be determined.

```
 600 yards;   divide by 14
 700 yards;   divide by 13
 800 yards;   divide by 13
 900 yards;   divide by 12
1000 yards;   divide by 11
```

There is another means of computing windage changes when increasing in range from 300 to 600 yards and 600 to 1000 yards. The sniper, when moving from 300 to 600 yards, simply doubles the amount of windage he used at 300 yards, adds one additional change and fires the shot. The same procedure is used when moving from 600 to 1000 yards, as a general rule, the shot will be within a minute of your point of aim. To adjust for ranges which fall between these three primary ranges, the sniper simply modifies his adjustment.

<u>TRANSITION</u>. The subject of light is controversial as light may have no effect on some snipers since light effects people in different ways.

2. <u>Effects of Light</u>.

a. Explanation of Light and It's Effects

(1) The light of day gives form and shape to everything the human eye can see. On a bright day, everything we see has height, width, and depth. In other words, three dimensions. The sun casts shadows on everything so that we get this feeling of depth

(2) Sometimes, the position of the sun will cast long shadows on a target, giving some of us an illusion of depth, where in fact, there is not any depth between your aiming point on the target and the target itself.

(3) When the sun is obscured by clouds, such as in an overcast sky, there are no shadows, everything seems to be flat, or in two dimensions.

(4) Each of us vary in the way we see things because we all have somewhat different eyesight. A worthwhile rule that can be applied is when the light is up or bright, the elevation may need to increase and the opposite applies for cloudy or dull light.

3. <u>Humidity and Temperature</u>.

a. <u>Humidity</u>. As humidity increases, the air density increases, offering more resistance to the flight of the bullet through it. This resistance slows the bullet and lowers it's point of impact, requiring the sniper to raise his elevation. The effects of humidity obviously increase with range. Changing combinations of weather cause variations in the effects on the bullet. For this reason, a sniper may fire two successive days on the same range and under what appears to be the same conditions and yet, require two different sight settings. For example, a rise in the humidity of 30 percent cannot always be noticed readily. This rise in humidity makes the air more dense. If this heavier air is present with a 10 mile per hour wind, it will require more elevation and more windage to hit the same location than when the humidity was 30% lower.

b. <u>Temperature</u>. The effect of temperature on the bullet is opposite to that of humidity, as higher temperatures of barral and powder will increase chamber pressure and, consequently, the muzzle velocity. This will cause the bullet to strike high and requires lowering the elevation. Tests have been conducted which indicate a variation in velocity of about 1.7 feet per second per degree of temperature, exclusive of the humidity factor. In such a case, 5 degrees would change the bullet strike by approximately 1 minute of elevation.

<u>OPPORTUNITY FOR QUESTIONS AND COMMENTS</u>

<u>SUMMARY</u>

1. <u>Reemphasize</u>. During this period of instruction, we have discussed the effects of wind, how it affects the flight of the bullet. We covered the various methods, step-by-step, used by the sniper to compute wind velocity and the subsequent formula used to compute the number of minutes of windage to be placed on the sight. We discussed mirage and where and how you look for mirage when you are in the field and how it will aid the sniper in determining wind direction and velocity. We then covered light and it's affect on the sniper as well as his target.

Humidity and Temperature were covered and their effects on the sniper as well as his ammunition. It is imperative that the sniper notices the change in humidity and temperature from day to day, as this will have an affect on his zero.

2. <u>Remotivate</u>. What would you consider one of the greatest aids to the sniper? Good Rifle? Position? Sight Picture? All of these are essential to a sniper, but even with all these aids, there is one more aid the sniper needs, knowledge of the effects of weather.

A good working knowledge of the effects of weather on you and your weapon will improve your ability to achieve that first round kill.

UNITED STATES MARINE CORPS
SCOUT/SNIPER INSTRUCTOR SCHOOL
Marksmanship Training Unit, Weapons Training Battalion
Marine Corps Development and Education Command
Quantico, Virginia 22134

SI0009
L, D, A

_____
(DATE)

INDIVIDUAL MOVEMENT

DETAILED OUTLINE

INTRODUCTION

1. **Gain Attention.** During the German occupation of Norway in W. W. II, what forces Norway had were overrun. This did not stop resistance, however. The Norwegians formed a strong underground to fight the Germans. During this struggle, there was a particular Norwegian soldier used as a sniper. Throughout the war, he is credited with killing over 300 German troops, and he was never caught or killed. How did he accomplish this phenominal feet, probably without even a telescope for his rifle? With properly selected and camouflaged firing positions, and proper use of tactical movement to stalk close to his targets. A sniper isn't often seen and therefore, not often shot at.

2. **Motivate.** This example shows that continued success at moving undetected is possible if the proper techniques of individual movement are applied with a high degree of skill. A sniper cannot move haphazardly. Since snipers move in small teams and are usually close to or behind enemy lines, their being spotted might mean capture or death for team members.

3. **State Purpose and Main Ideas**

   a. **Purpose.** The purpose of this period of instruction is to provide the student, as a potential sniper, the proper methods of tactical movement to remain undetected, and stalking into position for a shot.

   b. **Main Ideas.** The main ideas which will be discussed are the following:

      (1) Picking proper routes
      (2) Day and Night Movement
      (3) The Proper Methods of Crawling and Walking, and
      (4) Stalking.

4. **Learning Objectives.** Upon completion of this period of instruction, the student, will be able to:

   a. Apply the proper techniques of the low crawl, the high crawl, hand-and-knees crawl, and walking silently, day or night.

   b. Properly select routes to and from an objective or firing position.

c. Apply the proper methods of stalking to move undetected within a range of a probable target.

BODY

1. Preparation for Movement.

    a. General.

        (1) Sequence. Before a mission, there are a few items of preparation a sniper needs to pay particular attention to. One is to plan a primary and an alternate route to and from the objective. This is done by studying in-depth, large scale maps, aerial photographs of the area, and talking to people who have been through the areas before. Know as much about an area as possible before moving through it. Allow enough time for proper camouflage, which should be nearly perfect for the type of terrain to be moved through. Prior to move-ment, an inspection should be held for all personnel to insure that all shiny equipment is toned down, and all gear silenced. Also being sure only the min-imum amount of gear is taken that will be needed on the mission.

        (2) Route Selection. In selecting routes of movement, a sniper should try to avoid known enemy positions and obstacles. Open areas and exposed ridges should be avioded. Seek routes with cover and concealment, while trails should never be used. Take advantage of more difficult terrain; swamps, dense woods, etc. Avoid areas believed to be under enemy observation, mined or bobby trapped. Villages, or where you are likely to meet natives whould be skirted.

2. Day Movement

    a. General.

        (1) Normal infantry movement doesn't fully apply to a sniper. A sniper cannot afford to be seen at all. This means he has to be doubly careful, which in turn means he has to move considerably slower.

        (2) Two important rules for movement to remember are: Always assume your area is under observation, and secondly, during movement, stop, look, and listen, plan your route, then move by bounds.

        (3) Always observe from a covered position, as low to the ground as possible. Look around objects or through brush, not over it, noting in detail everything, and using binoculars if needed. Also, blend into the background such as grass or bushes before observing. Listen to every sound. Your senses must be fully alert. Plan your route to your next observation point. Move under the most concealed routes, by using necessary methods of walking or crawling. Once to your next point, repeat the process. The type of terrain will dictate the speed of travel. It may mean moving slow, but if you are spotted, your life and mission are compromised.

    b. Types of Movement.

        (1) Walking. Wherever you are walking, walk carefully, distinctly, and quietly. Be concious of every step you take.

(a) Walk at a crouch to maintain a low profile with shadows and bushes. Most enemy will be looking for an upright man.

(b) Very slowly lift one foot and move it forward, clearing obstacles with the toes straight to the front.

(c) Pick out a point about ½ a normal stride in front, preferably free of dry leaves and twigs, for silence, then place the toes or outside edge of the foot lightly down to get the feel of the ground. Rotate the foot down onto the ball of the foot. Continue placing the foot until the heel is down.

(d) Now, very slowly, start shifting the body weight forward until it all rests on the forward foot, but slowly enough that it makes no sound.

(e) Then, with the opposite foot, repeat the process. Again, the terrain will determine the speed and silence of movement.

(2) Low Crawl. The low crawl is used when cover and concealment are low or scarce, when the enemy is near or has a clear field of view to your position. It is slow, so speed cannot be essential.

(a) First, lay the body flat on the ground as possible, legs together, inside ankles on the deck, arms to the front and flat on the ground.

(b) Grasp the rifle by the upper sling swivel, letting the stock rest on your forearm and the butt resting on the ground. This carry is to protect the rifle from abuse. Be sure the muzzle doesn't protrude into the air.

(c) To move forward, extend the arms fully, to the front, and pull one leg forward. Then very slowly, pull the arms and push with the leg.

(d) When the pushing leg is tired, the opposite leg can take over, but use only one leg for a sequence of pushing. All movement must be very slow and deliberate. All parts of the body are kept as low to the ground as possible, especially the head.

(3) High Crawl. The high crawl is used when cover is more prevalent, or more speed is required.

(a) The body is kept free of the ground, and your weight is resting on your arms and lower legs. The rifle is either carried as in the low crawl or cradled in the arms.

(b) Movement is made by alternately pulling with each arm, and pushing with one leg if one still wishes to remain fairly low, or alternating legs for pushing if there is adequate cover.

(c) Be concious so as not to allow the head and buttocks to raise too high, and the legs making excessive noise being dragged over brush and debris.

(4) _Turning While Crawling_. If it is necessary to change direction or turn completely while withdrawing, when crawling:

(a) First, when extreme care is needed, ease the body as far to the right as possible, but keep the legs together. The left leg is then moved as far to the left as possible and the right leg then closed to it. This will effect a turn to the right and should be repeated until the sniper is facing the required direction. The procedure should be reversed to effect a move to the left.

(b) _Moving Backwards_. It may be sometimes necessary to withdraw without turning, this can be done by the low crawl in reverse, pushing instead of pulling with the arms.

(5) _Hands and Knees Crawl_. When cover is adequate, or silence is necessary, crawling on hands and knees can be used.

(a) The rifle is held in one hand close to the chest and in line with the body. The sling is grasped with the stock to keep it from being tangled on the ground. The weight of the upper body is supported by the opposite arm.

(b) Supporting the rifle in the left hand, pick a point ahead to position the right hand. Move it slowly into position making no noise. While moving the right arm, the weight of the upper body can be supported by leaning on the left elbow. Once the right arm is placed, the left arm and rifle is moved forward.

(c) A position is then picked to move the knees to. Then in turn, each leg is lifted to clear any obstruction and softly placed into it's new position. Again, the situation, ground cover, and terrain will determine the speed and silence of movement.

(d) If absolute silence is needed, leaves, twigs, and pebbles can be removed before placing the hands and knees. The movement must be very, very slow, and soft. Breathing should be shallow through the mouth, if very close to the enemy. Do not look directly at him.

3. _Night Movement_.

a. _General_. Night movement is essentially the same as in the day, but of course, visibility is greatly limited. One has to rely on the senses of touch and hearing to a greater extent. A sniper should move, if at all possible, under the cover of darkness, fog, or haze to conceal his movements. This is a good safety factor, but hidden enemy are harder to spot, and specific positions or landmarks are harder to locate. Before moving at night, let your eyes adjust to the darkness for at least 30 minutes. To distinguish an object in the dark, get low to the ground, to silhouette it against the sky, also look 5 - 10 degrees away from the object. If one looks directly at an object in the darkness, it will distort, or when the eyes are tired, completely disappear. Concealment is not as critical at night, but staying next to a dark background, and not being silhouetted is. Quick movement at night is easily seen, and sound travels further and clearer.

Night Movement Continued

So, in the darkness of night, slow and silent movement is essential. While moving, listen to the night time noises for anything out of place or unusual, and continually scan for movement. Also, take advantage of wind and other noises to mask your own movements.

    b. **Types of Movement.** Movement at night is essentially the same as in the day, except it must be more deliberate and done by feel.

        (1) **Walking.** Walking at night is basically the same as in the day.

            (a) The arm not carrying the rifle is used as a feeler, a guide against obstacles, trip wires, and moving through brush. The arm is also kept in front to stop the body in case of falling.

            (b) The ground is softly felt by the toe before placement of the foot, so absolute silence is maintained. The feet are raised high to clear any brush or other obstacles.

        (2) **Low Crawl.** The low crawl is the same at night, except the body is lifted off the ground by the forearms, leg or toes, just enough so that when moving, no shuffle noise is made. The hands are used as feelers to determine the way.

        (3) **High Crawl.** The amendments to the high crawl would be the same as for the low crawl at night.

        (4) **Hands and Knees Crawl.** Again, this is the same as in the day, except slower movement is needed.

            (a) The free hand picks out a place for each knee to move to and helps guide it there, while moving feel from head height down, for obstacles and trip wires.

        (5) **Additional Information.** At night, the senses have to be relied upon to a great extent, learn to trust them and be able to interpret what they are telling you. The enemy may even be located by the sense of smell. Also, your gear whould be arranged so that it may be gotten quickly in the dark, and always keep it within hands reach.

4. **Stalking.**   a. The object of stalking is that the sniper should move, unseen, to a fire position within such range of his target that he is sure of a first round kill, and then to withdraw unseen. The practical art of stalking incorporates the application of all aspects of fieldcraft, and is such that it can only be effectively learnt by repeated practice, and discussion of practice over various types of ground.

    b. **Reconnaissance.** Any stalk undertaken without first doing a thorough reconnaissance is likely to have limited success. Opportunities to view the ground, though desirable, will be rare in battle, so the sniper must be an expert with the map and aerial photograph so that the maximum information can be gained from both.

c. **Before Stalking.**

(1) The exact location of the enemy position to be stalked should be noted and memorized. Particular attention should be given to nearby features and landmarks.

(2) Decide upon an area which appears to present the best possible fire position, though the exact fire position can rarely be pinpointed in advance.

(3) Select the best line of advance and split into bounds; each bound can then be considered in greater detail as it is arrived at on the ground. Remember, once a sniper is committed to a line of advance, he may find great difficulty in changing it, so great skill is needed in the initial appreciation. Particular points to consider are:

(a) The availability of natural cover, and in particular, any dead ground.

(b) The position, and frequency of any obstacles, whether natural or artificial.

(c) Likely points along the line of advance from which observations can be made. When possible, these should coincide with the finish and start of the planned bounds.

(d) The location of any other known or possible enemy locations.

(e) The general method of movement likely to be possible for each bound (eg, crawling or walking). This appreciation is important since it will be this in relation to the distance to be stalked that will dictate the length of time required.

(f) The withdrawal route should differ from that of the approach if it is at all possible, but should be planned in a similar manner. It is important that patience is maintained during a withdrawal, since the enemy will be much more alert at this stage than during the approach.

d. **While Stalking.**

(1) It is easy to lose the sense of direction when stalking, particularly if the sniper has to crawl for any appreciable distance. The chances of this happening can be reduced if:

(a) The use of the compass, map and air photograph have been thoroughly mastered.

(b) A distinct landmark or two, or even a series, can be memorized.

(c) The direction of the wind and sun are noted, but bear in mind that over a long period of time the wind direction can change, and the sun will change position.

(2) The sniper must be alert at all times. Any relaxation on a stalk could lead to carelessness, resulting in an unsuccessful mission.

(3) Observation must be done with care, and at frequent intervals. It is particularly important at the beginning and end of each bound.

(4) If suprised or exposed during a stalk, instinctive immediate reaction is necessary. The sniper must decide whether to freeze and remain immobile, or to move quickly to the nearest cover away from the point of exposure.

(5) Remember, disturbed animals and birds can draw attention to the area of the approach if not the exact position.

(6) Take advantage of any local disturbances or distractions that may enable quicker movement than would otherwise have been possible. It should be emphasized though that such movement involves a degree of risk, and when the enemy is close, risks should be avoided.

(7) Keep in mind any changes in local cover since such changes will usually require an alteration to personal camouflage.

e. Stalking at Night.

(1) It may be necessary for the sniper to stalk at night, possibly in order to occupy an OP or a fire position under the cover of darkness. The problems are much the same as stalking in daylight, except that a man is less well adapted for movement at night. The principle differences are:

(a) There is a degree of protection offered by the darkness against aimed enemy fire.

(b) Whilst observation is still important, much more use is made of hearing, and in consequence silence is vital.

(c) Cover is less important than background, and in particular, crests and skylines should be avoided.

(d) Maintenance of direction is much more difficult to achieve, and places greater emphasis on thorough reconnaissance. A compass or a knowledge of the stars may be of assistance.

(2) The Starlight Scope is extremely useful when stalking at night and should be used off the weapon as an observation aid in the early stages of the stalk.

OPPORTUNITY FOR QUESTIONS AND COMMENTS

SUMMARY

1. <u>Reemphasize</u>. In this period of instruction we have covered:

    a. A few ways of preparation for a sniper mission. One of the most important ways to do an indepth study of the area of movement. Also, we covered proper route selection, so the sniper has the best chance of moving unobserved.

    b. We learned two important rules for movement which are, always assume your area is under observation, and during movement, stop, look, and listen often, plan a route, then move by bounds.

    c. Next were the types of movement, which are: Walking, low crawl, high crawl, hands and knees crawl, and how they were to be used day or night.

2. <u>Remotivate</u>. A person can be taught to move quiet enough and undetected to such degree that he can be able to stalk directly up to a man or animal in almost any type of terrain. Once the proper technique is taught, there must be persistence, patience, attention to detail, and the willingness to learn. That students, must come from you if you are to become the type of snipers we would have you to be. That would be to be able to make a kill anywhere in the world, and leave no indication that you were even there.

UNITED STATES MARINE CORPS
SCOUT/SNIPER INSTRUCTOR SCHOOL
Marksmanship Training Unit, Weapons Training Battalion
Marine Corps Development and Education Command
Quantico, Virginia 22134

SI0010
L, D, A

_____
(DATE)

RANGE ESTIMATION TECHNIQUES

DETAILED OUTLINE

INTRODUCTION

1. Gain Attention. Everyone has had to estimate the distance from one point
to another at some time. Usually an estimate was made either because no tool
was available for exact measurements or because time did not allow such a
measurement.

2. Motivate. As a sniper, in order to engage a target accurately, you will
be required to estimate the range to that target. However, unlike with many
of your early experiences, a "ball-park guesstimate" will no longer suffice.
You will have to be able to estimate ranges out to 1000 yards with 90%
accuracy.

3. State Purpose and Main Ideas.

    a. Purpose. To acquaint the student with the various techniques of range
estimation he will use in his role as a sniper.

    b. Main Ideas. Describe the use of maps, the 100 meter method, the
appearance of objects method, the bracketing method, the averaging method, the
range card method and the use of the scope reticle in determining range.

4. Learning Objectives. Upon completion of this period of instruction, the
student will:

    a. Determine range with the aid of a map.

    b. Demonstrate the other techniques for determining range by eye.

    c. Identify those factors which effect range estimation.

TRANSITION. The sniper's training must concentrate on methods which are
adaptable to the sniper's equipment and which will not expose the sniper.

BODY   A.   METHODS OF RANGE ESTIMATION

1. Use of Maps. When available, maps are the most accurate aid in determining range. This is easily done by using the paper-strip method for measuring horizontal distance.

2. The 100 Meter Unit of Measure Method.

   a. Techniques. To use this method, the sniper must be able to visualize a distance of 100 meters on the ground. For ranges up to 500 meters, he determines the number of 100 meter increments between the two points. Beyond 500 meters, he selects a point midway to the targets, determines the number of 100 meter increments to the halfway point, and doubles the result.

   b. Ground which slopes upward gives the illusion of greater distance, while ground sloping downward gives an illusion of shorter than actual distance.

   c. Attaining Proficiency. To become proficient with this method of range estimation, the sniper must measure off several 100 meter courses on different types of terrain, and then, by walking over these courses several times, determines the average number of paces required to cover the 100m of the various terrains. He can then practice estimation by walking over unmeasured terrain, counting his paces, and marking off 100m increments. Looking back over his trail, he can study the appearance of the successive increments. Conversely, he can estimate the distance to a given point, walk to it counting his paces, and thus check his accuracy.

   d. Limitations. The greatest limitation to the 100m unit of measure method is that it's accuracy is directly related to how much of the terrain is visible to the observer. This is particularly important in estimating long ranges. If a target appears at a range of 100 meters or more, and the observer can only see a portion of the ground between himself and the target, the 100m unit of measure method cannot be used with any degree of accuracy.

3. Appearance-of-Objects Method.

   a. Techniques. This method is a means of determining range by the size and other characteristic details of some object. For example, a motorist is not interested in exact distances, but only that he has sufficient road space to pass the car in front of him safely. Suppose however, that a motorist knew that at a distance of 1 kilometer (Km), an oncoming vehicle appeared to be 1 inch high, 2 inches wide, with about ½ inch between the headlights. Then, any time he saw oncoming vehicles that fit these dimensions, he would know that they were about 1 Km away. This same technique can be used by snipers to determine range. Aware of the sizes and details of personnel and equipment at known ranges, he can compare these characteristics to similar objects at unknown distances, and thus estimate the range.

   b. To use the appearance-of-objects method with any degree of accuracy, the sniper must be thoroughly familiar with the characteristic details of objects as they appear at various ranges. For example, the sniper should study the appearance of a man at a range of 100 meters. He fixes the man's appearance firmly in his mind, carefully noting details of size and the characteristics of uniform and equipment. Next, he studies the same man in the kneeling position and then in the prone position. By comparing the appearance of these positions at known ranges from 100-500m, the sniper can establish a series of mental images which will help him in determining ranges on unfamiliar terrain. Practice time should also be devoted to the appearance of other familiar objects wuch as weapons and vehicles.

   c. **Limitations.** Because the successful use of this method depends upon visibility, or anything which limits visibility, such as smoke, weather or darkness, will also limit the effectiveness of this method.

4. **Combination of Methods.** Under proper conditions, either the 100m unit of measure or the appearance-of-objects method of determining range will work, however, proper conditions rarely exist on the battlefield. Consequently, the sniper will be required to use a combination of methods. Terrain can limit the accuracy of the 100m unit of measure method and visibility can limit the appearance-of-objects method. For example, an observer may not be able to see all of the terrain out to the target, but he may see enough to get a fair idea of the distance. A slight haze may obscure many of the target details, but the observer can still make some judgment of it's size. Thus, by carefully considering the results of both methods, an experienced observer should arrive at a figure close to the true range.

5. **Bracketing Method.** By this method, the sniper assumes that the target is no more than "X" meters, but no less than "Y" meters away; he uses the average as the estimation of range.

6. **Averaging Method.** Snipers can increase the accuracy of range estimation by eye by using an average of the individual team members estimations.

7. **Range Card Method.** Information contained on prepared range cards establishes reference points from which the sniper can judge ranges rapidly and accurately. When a target appears, it's position is determined in relation to the nearest object or terrain feature drawn on the range card. This will give an approximation of the targets range. The sniper determines the difference in range between the reference point and the target, and sets his sights for the proper range, or uses the correct hold off.

8. **Range Estimation Formula Method.** This method requires the use of either binoculars or telescopic sights equipped with a mil scale. To use the formula, the sniper must know the average size of a man or any given piece of equipment and he must be able to express the height of the target in yards.
The formula is: $\dfrac{\text{SIZE OF OBJ. (IN YDS) X 1000}}{\text{SIZE OF OBJ. (IN MILS)}} = \text{RANGE TO TARGET}$

For example: A sniper, looking through his scope sees a man standing. He measures the size of the man, using the mil scale on the reticle, and he sees that the man is 4 mils high. He knows that the average man is five and a half feet tall. To convert 6 feet to yards, he divides by 3 and finds that the man is 2.0 yards tall. Using the Formula:

$$\frac{\text{SIZE OF OBJ (IN YDS)}}{\text{SIZE OF OBJ (IN MILS)}} \frac{2.0 \times 1000}{4} = \frac{2000}{4} = 500 \text{ yards}$$

Once the formula is understood, the sniper needs only to be able to estimate the actual height of any target and he can determine the range to that target extremely accurately.

   b. _Limitations._ While this formula can be extremely accurate, it does have several limitations.

      (1) At long ranges, measurement in mils must be precise to the nearest half mil or a miss will result. For example; If a man standing appears to be 1½mils high, he is 1333 yds away. If he appears to be 2 mils high, he is only1000 yds away. Careless measurement could result in a range estimation error of 333 yds in this case.

      (2) This formula can be worked quickly, even if the computations are done mentally. However, as with any formula, care must be taken in working it or a totally wrong answer can result, and

      (3) The formula depends entirely on the sniper's ability to estimate the actual height of a target in yards.

B. FACTORS EFFECTING RANGE ESTIMATION

1. _Nature of the Target_

   a. An object of regular outline, such as a house, will appear closer than one of irregular outline, such as a clump of trees.

   b. A target which contrasts with it's background will appear to be closer than it actually is.

   c. A partially exposed target will appear more distant than it actually is.

2. _Nature of Terrain._ The observer's eye follows the irregularities of terrain conformation, and he will tend to overestimate distance values. In observing over smooth terrain such as sand, water, or snow, his tendency is to underestimate.

3. _Light Conditions._ The more clearly a target can be seen, the closer it appears. A target in full sunlight appears to be closer than the same target when viewed at dusk or dawn, through smoke, fog or rain. The position of the sun in relation to the target also affects the apparent range. When the sun is behind the viewer, the target appears to be closer. When the sun is behind the target, the target is more difficult to see, and appears to be farther away.

OPPORTUNITY FOR QUESTIONS AND COMMENTS

SUMMARY

1. **Reemphasize.** We have seen various ways to estimate range. Each one of them works well under the conditions for which it was devised, and, when used in combination with one another, will suit any condition of visibility or terrain.

2. **Remotivate.** The accuracy of the shot you will fire will depend to a large extent on whether or not you have applied the rules for range estimation. Remember, if you cannot determine how far your target is away from you, you would just as well have left your rifle in the armory.

RANGE ESTIMATION TABLE FOR SIX FOOT MAN

Average Standing Man - 6 Feet Tall/2 Yards Tall

Average Sitting/Kneeling Man - 3 Feet Tall/1 Yard Tall

HEIGHT IN MILS	STANDING RANGE	SITTING/KNEELING RANGE
1	2000	1000
1.5	1333	666
2	1000	500
2.5	800	400
3	666	333
3.5	571	286
4	500	250
4.5	444	222
5	400	200
5.5	364	182
6	333	167
6.5	308	154
7	286	143
7.5	267	133
8	250	125
8.5	235	118
9	222	111
9.5	211	105
10	200	100

UNITED STATES MARINE CORPS
SCOUT/SNIPER INSTRUCTOR SCHOOL
Marksmanship Training Unit, Weapons Training Battalion
Marine Corps Development and Education Command
Quantico, Virginia 22134

SI 0011
L, D, A

_____
(DATE)

TECHNIQUES OF OBSERVATION

DETAILED OUTLINE

INTRODUCTION

1. <u>Gain Attention</u>. The sniper's mission requires him to support combat operations by delivering precision fire from concealed positions to selected targets. The term "selected targets" correctly implies that the sniper is more concerned with the significance of his targets than with the number of them. In his process of observation, he will not shoot the first one available, but will index the location and identification of all the targets he can observe.

2. <u>Motivate</u>. The sniper is expected to perform several missions other than sniping. One of the more important is observation of the enemy and his activities.

3. <u>State Purpose and Main Ideas</u>.

    a. <u>Purpose</u>. The purpose of this period of instruction is to provide the student with the knowledge, procedures and techniques applicable to both day and night time observation.

    b. <u>Main Ideas</u>. The main ideas which will be discussed are the following:

        (1)   Observation Capabilities and Limitations
        (2)   Observation Procedures

4. <u>Learning Objectives</u>. At the conclusion of this period of instruction, the student, without the aid of references, will be able to:

    a. Describe the limitations of observation and the steps to be taken to overcome those limitations.

    b. Describe the use of the telescope, 3 x 9 variable, and the starlight scope as an observation aid.

    c. Describe the procedures used to observe and maintain observation of a specific area or target.

TRANSITION. Observation is the keynote to a sniper's success. He must be fully aware of the human capabilities and limitations for productive observation in waning light and in darkness, and of his aids, which can enhance his visual powers under those conditions.

BODY

1. Capabilities and Limitations.

   a. Night Vision. Night runs the gamut from absolute darkness to bright moonlight. No matter how bright the night may appear to be, however, it will never permit the human eye to function with daylight precision. For maximum effectiveness, the sniper must apply the proven principles of night vision.

      (1) Darkness Adaptation. It takes the eye about 30 minutes to regulate itself to a marked lowering of illumination. During that time, the pupils are expanding and the eyes are not reliable. In instances when the sniper is to depart on a mission during darkness, it is recommended that he wear red glasses while in lighted areas prior to his departure.

      (2) Off-Center vs. Direct Vision. (TA #1) Off-center vision is the technique of focusing attention on an object without looking directly at it. An object under direct gaze in dim light will blur and appear to change shape, fade, and reappear in still another form. If the eyes are focused at different points around the object and about 6 to 10 degrees away from it, side vision will provide a true picture of the object.

      (3) Scanning. Scanning is the act of moving the eyes in short, abrupt, irregular changes of focus around the object of interest. The eye must stop momentarily at each point, of course, since it cannot see while moving.

      (4) Factors Affecting Night Vision

         (a) Lack of vitamin A impairs night vision. However, overdoses of vitamin A will not improve night vision.

         (b) Colds, headache, fatigue, narcotics, heavy smoking, and alcohol excess all reduce night vision.

         (c) Exposure to a bright light impairs night vision and necessitates a readaptation to darkness.

         (d) Darkness blots out detail. The sniper must learn to recognize objects and persons from outline alone.

   b. Twilight. During dawn and dusk, the constantly changing natural light level causes an equally constant process of eye adjustment. During these periods, the sniper must be especially alert to the treachery of half light and shadow. Twilight induces a false sense of security, and the sniper must be doubly careful for his own safety. For the same reason, the enemy is prone to carelessness and will frequently expose himself to the watchful sniper. The crosshairs of the telescopic sight are visible from about one-half hour prior to sunrise until about one-half houw after sunset.

c. **Illumination Aids.** On occasion, the sniper may have the assistance of artificial illumination for observation and firing.

EXAMPLES:

(1) **Cartridge, Illuminating, M301A2.** Fired from an 81mm mortar, this shell produces 50,000 candlepower of light which is sufficient for use of the binoculars, the M49 spotting scope, or the rifle telescopic sight.

(2) **Searchlights.** In an area illuminated by searchlight, the sniper can use any of the above equipment with excellent advantage.

(3) **Other.** Enemy campfires or lighted areas and buildings are other aids to the observing sniper.

d. **Observation Aids.**

(1) **Binoculars.** (TA #2) Of the night observation aids, binoculars are the simplest and fastest to use. They are easily manipulated and the scope of coverage is limited only by the sniper's scanning ability. Each sniper team will be equipped with binoculars to aid in observing the enemy and in searching for and selecting targets. The binocular, M17A1, 7 x 50, has seven power magnification and a 50mm objective lens. Focal adjustments are on the eyepiece with separate adjustments for each eye. The left monocle has a horizontal and vertical scale pattern graduated in mils that is visible when the binoculars are in use.

(a) **Method of Holding.** (TA # 3) Binoculars should be held lightly, monocles resting on and supported by the heels of the hands. The thumbs block out light that would enter between the eye and the eyepiece. (TA #4) The eyepieces are held lightly to the eye to aviod transmission of body movement. Whenever possible, a stationary rest should support the elbows.

(b) **Adjustments**

(1) **Interpupillary Adjustment.** The interpupillary distance (distance between the eyes) varies with individuals. The two monocles that make up a pair of field glasses are hinged together so that the receptive lenses can be centered over the pupils of the eyes. Most binoculars have a scale on the hinge, allowing the sniper to preset the glasses for interpupillary distance. To determine this setting, the hinge is adjusted until the field of vision ceases to be two overlapping circles and appears as a single sharply defined circle. (TA # 5)

(2) **Focal Adjustment.** Each individual and each eye of that individual requires different focus settings. Adjust the focus for each eye in the following manner:

(a) With both eyes open, look through the glasses at a distant object.

(b) Place one hand over the objective lens of the right monocle and turn the focusing ring to the left monocle until the object is sharply defined.

(c) Uncover the right monocle and cover the left one.

(d) Rotate the focusing ring of the right monocle until the object is sharply defined.

(e) Uncover the left monocle; the object should then be clear to both eyes.

(f) Read the diopter scale on each focusing ring and record for future reference.

(c) <u>Reticle</u>. (TA #6) The mil scale that is etched into the left lens of the binoculars is the reticle pattern and is used in adjusting artillery fire and measuring vertical distance in mils. The horizontal scale is divided into 10-mil increments. The zero line is the short vertical line that projects below the horizontal scale between two numbers "1". To measure the angle between two objects (such as a target and an artillery burst), center the target above the zero line. Then read the number which appears on the scale under the artillery burst. There are two sets of mil scales, one above the zero on the horizontal scale, the other above the left horizontal 50-mil line on the horizontal scale. The vertical scales are divided into increments of 5 mils each. The vertical angle between the house and point A at the base of the tree is 10 mils. The third vertical scale is the range scale. It is used to estimate ranges from a known range but is not used by the sniper since he estimates his ranges by eye.

(2) <u>Rifle Telescopic Sight, 3 x 9 Variable</u>. When equipped with the telescopic sight, the sniper can observe up to 800 meters with varying effectiveness in artificial illumination. In full moonlight, it is effective up to 600 meters. For best results, a supported position should be used. The power of the telescopic sight may be adjusted to the situation. However, the lower the power, the greater is the lightgathering quality of the telescopic sight. Optional power adjustments are as follows:

(a) <u>3 to 5 Power (Low)</u>. This power gives a wide field of view, but the objects viewed will appear small. Visibility is good to 600 meters, and the crosshairs and range scale are well defined.

(b) <u>6 and 7 Power (Medium)</u>. This offers the best observation character. Field of view is reduced, but objects are more discernable and clarity is increased. Crosshairs, range scale, and reference wire are clear enough for shooting.

(c) <u>8 and 9 Power (High)</u>. At high power, the field of view is more reduced and scanning clarity is impaired. High power can be used to distinguish specific objects, but scanning will lend a flat, unfocused appearance to terrain.

(3) _Starlight Scope_. (TA #7) Although the function of the star-light scope is to provide an efficient viewing capability during the conduct of night combat operations, the starlight scope does not give the width, depth, or clarity of daylight vision. However, the individual can see well enough at night to aim and fire his weapon, to observe effect of firing, the terrain, the enemy, and his own forces; and to perform numerous other tasks that confront a Marine in night combat. The starlight scope may be used by snipers to:

(a) Assist sniper teams in deployment under cover of darkness to preselected positions.

(b) Assist sniper teams to move undetected to alternate positions.

(c) Locate and suppress hostile fire.

(d) Limit or deny the enemy movement at night.

(e) Counter enemy sniper fire.

a. _Factors Affecting Employment_. Consideration of the factors affecting employment and proper use of the starlight scope will permit more effective execution of night operations. The degree to which these factors aid or limit the operational capabilities of the starlight scope will vary depending on the light level, weather conditions, operator eye fatigue, and terrain over which the starlight scope is being employed.

(1) _Light_. Since the starlight scope is designed to function using the ambient light of the night sky, the most effective operation can be expected under conditions of bright moonlight and starlight. As the ambient light level decreases, the viewing capabilities of the starlight scope diminish. When the sky is overcast and the ambient light level is low, the viewing capabilities of the starlight scope can be greatly increased by the use of flares, illuminating shells or searchlight.

(2) _Weather Conditions_. Clear nights provide the most favorable operating conditions in that sleet, snow, smoke, or fog affect the viewing capabilities of the starlight scope. Even so, the starlight scope can be expected to provide some degree or viewing capability in adverse weather conditions.

(3) _Terrain_. Different terrain will have an adverse effect on the starlight scope due to the varying ambient light conditions which exist. It will be the sniper's responsibility to evaluate these conditions and know how each will affect his ability to observe and shoot.

(4) _Eye Fatigue_. Most operators will initially experience eye fatigue after five or ten minutes of continuous observation through the starlight scope. To aid in maintaining a continued viewing capability and lessen eye fatigue, the operator may alternate eyes during the viewing period.

4. _M49 Observation Telescope_. (TA # 8) The M49 observation telescope is a prismatic optical instrument of 20-power magnification. It is carried by the

sniper teams whenever justified by the nature of a mission. The lens of the telescope are coated with a hard film of magnesium flouride for maximum light transmission. This coating together with the high magnification of the telescope makes observation and target detection possible when conditions or situations would otherwise prevent positive target identification. Camouflaged targets and those in deep shadows can be distinguished, troop movements can be observed at great distances, and selective targets can be identified more readily.

a. Operation. The eyepiece cover cap and objective lens cover must be unscrewed and removed from the telescope before it can be used. The cap and cover protect the optics when the telescope is not in use. The eyepiece focusing sleeve is turned clockwise or counterclockwise until the image can be carefully seen by the operator. CAUTION: Care must be taken to prevent cross-threading of the fine threads.

(2) Observation Procedures. The sniper, having settled into the best obtainable position, is ready to search his chosen area. The process of observation is planned and systematic. His first consideration is towards the discovery of any immediate danger to himself, so he begins with a "hasty search" of the entire area. This is followed by a slow, deliberate observation which he calls a "detailed search". Then, as long as he remains in position, the sniper maintains a constant observation of the area using the hasty and detailed search methods as the occasion requires.

(a) Hasty Search. This is a very rapid check for enemy activity conducted by both the sniper and the observer. The observer makes the search with the 7 x 50 binoculars, making quick glances at specific points throughout the area, not by a sweep of the terrain in one continuous panoramic view. The 7 x 50 binoculars are used in this type search because they afford the observer with the wide field of view necessary to cover a large area in a short time. The hasty search is effective because the eyes are sensitive to any slight movements occurring within a wide arc of the object upon which they are focused. The sniper, when conducting his hasty search, uses this faculty called "side vision" or "seeing out of the corner of the eye". The eyes must be focused on a specific point in order to have this sensitivity.

(b) Detailed Search.

(1) If the sniper and his partner fail to locate the enemy during the hasty search, they must then begin a systematic examination known as the 50-meter overlapping strip method of search. (TA # 8) Again the observer conducts this search with the 7 x 50 binoculars, affording him the widest view available. Normally, the area nearest the sniper offers the greatest potential danger to him. Therefore, the search should begin with the terrain nearest the oberserver's location. Beginning at either flank, the observer should systematically search the terrain to his front in a 180-degree arc, 50 meters in depth. After reaching the opposite flank, the observer should search the next area nearest his position. This search should cover the terrain located between approximately 40 and 90 meters of his position. The second search of the terrain includes about ten meters of the area examined during the first search. This technique ensures complete coverage of the area. Only when a target appears does the observer use the M49 observation scope to get a more detailed and precise description of the target. The observation scope should not be used to conduct either the hasty or detailed search as it limits the observer with such a small field of view. The observer continues searching from one flank to the other in 50-meter overlapping strips as far out as he can see.

(2) To again take advantage of his side vision, the observer should focus his eyes on specific points as he searches from one flank to the other. He should make mental notes of prominent terrain features and areas that may offer cover and/or concealment to the enemy. In this way, he becomes familiar with the terrain as he searches it.

(c) Maintaining Observation.

(1) Method. After completing his detailed search, the sniper may be required to maintain observation of the area. To do this, he should use a method similar to his hasty search of the area. That is, he uses quick glances at various points throughout the entire area, focusing his eyes on specific features.

(2) Sequence. In maintaining observation of the area, he should devise a set sequence of searching to ensure coverage of all terrain. Since it is entirely possible that this hasty search may fail to detect the initial movement of an enemy, the observer should periodically repeat a detailed search. A detailed search should also be conducted any time the attention of the observer is distracted.

OPPORTUNITY FOR QUESTIONS

SUMMARY

1. Reemphasize. During this period of instruction, we have discussed the capabilities and limitations of observation during the hours of both daylight and dark. The different techniques and aids of improving your vision were discussed.

We covered the night observation aids that are available to the sniper. It was noted that the binoculars are the simplest and fastest to use. The starlight scope was discussed in detail as to it's employment and those factors affecting it's employment.

In conclusion, details of observation procedures were covered. The "Hasty Search" is understood to be the first search conducted by the sniper once he moves into position as this search is conducted to discover any immediate danger to him.

2. Remotivate. Your ability to become proficient in the techniques mentioned will allow you to see the enemy before he sees you and get that first round off.

UNITED STATES MARINE CORPS
SCOUT/SNIPER INSTRUCTOR SCHOOL
Marksmanship Training Unit, Weapons Training Battalion
Marine Corps Development and Education Command
Quantico, Virginia 22134

SI0012
L, D, A

TARGET DETECTION AND SELECTION _____
                                      (DATE)

DETAILED OUTLINE

INTRODUCTION

1. **Gain Attention**. In Vietnam, a Marine Sniper once took three days to crawl 800 yards, fired one shot to complete his mission and started on his way back. He didn't stay in position longer to accumulate more kills, because he was sent on a specific mission, to take out one selected target. Today we are going to cover how to detect and select such key targets.

2. **Motivate**. The Marine rifleman can engage and kill the average enemy soldier, but a sniper doing this does not use his full potential. Killing a private will not nearly effect the situation as compared to killing a high ranking officer, that would disrupt command and communications of a large unit. The sniper is able to select his targets with care so that it will do the greatest possible damage to their moral and fighting ability.

3. **State Purpose and Main Ideas**.

   a. **Purpose**. The purpose of this class is to show the prospective sniper, how to select a proper target, how to detect such targets, and the actions leading up to the actual firing at a target.

   b. **Main Ideas**. The main ideas which will be discussed are the following:

      (1) Indexing Targets,
      (2) Detecting Targets, and
      (3) Selecting Targets.

4. **Learning Objectives**. At the end of this period of instruction, the student will:

   a. Understand the importance of proper target indexing and considerations in being able to engage targets accurately.

   b. Know what indicators to look for when trying to locate a target.

   c. Know how to select the most advantageous target and how to use that target as bait to bring more personnel under fire.

BODY

1. **General**. The sniper's mission requires him to support combat operations by delivering precision fire from concealed positions at selected targets. The term "selected targets" correctly implies that the sniper is more concerned with the significance of his targets than the number of them. In his process of target detection, he will not shoot the first one available, but will index the location and identification of all the targets he can observe. He will then fire at them in descending order of their importance.

Before covering the detecting and selecting of targets, we must touch on the aspects leading up to the detecting phase, to show that it is not done in a haphazard manner.

2. **Sniping Position**. Though it is possible to come across unexpected targets or targets of opportunity while on the move, the sniper should not rely on these as a primary sources for kills. The proper method is to select a specific area for observation, move to the area under the cover of darkness and set in a well concealed position with adequate cover. The position should have good fields of observation and fire, with a prearranged escape route, making sure of security to the rear to spot enemy who may stumble onto you from behind.

   a. **Searching**. When the sniper team is secure in position or at first light, a hasty search is done to detect any enemy in the immediate area, who may be dangerous to it. Once this is done, a detailed search is begun over the entire area of observation. This is where the art of observation comes into play, in that every minute object is studied and identified for possible evidence of the enemy.

   b. **Indexing Targets**. The sniper and observer working as a team must have an accurate method of relaying word of position of a possible target to one another. Consider the following conversation between the observer and sniper.
       O - "I see something over there." ----- S -"Over where?"
       O - "Way over there to the right." ---- S - "Where to the right?"
       O - "Beside that big tree." ---- S - "Which tree?"
   It is quick to see that a sniper team would not be effective unless there were a proper procedure for indexing targets by range, so they may be engaged quickly if need be. Also, the sniper may sight several targets at the same time and he will need to remember all their locations, while he determines which to engage first.

       (1) Prominent objects or terrain features are drawn on the sniper's range card, with their individual distances marked. When a possible target is spotted, it's location and range can be quickly noted by it's relation to one of the prominent features on the card. The log book can be a quick reference for information on previously sighted targets.

       (2) There are several methods of indicating the position of possible targets from others quickly and with no confusion. One method is to use the mil scale in the binoculars to give the location of an object from a known reference point. A hasty method of relating the position of a target is to use the width of the hand, fist, or fingers held at arms length. Such as "Three fingers left of the dead tree" or "One hand's width right of the building in the shrub bush."

   c. **Considerations**

(1) <u>Exposure Time</u>. Moving targets may expose themselves for only a short time. The sniper must be very alert to note the points of disappearance of as many as possible before he engages any one of them. By so doing, he will be able to take several of them under fire in rapid succession.

(2) <u>Number of Targets</u>. When the number of targets is such that the sniper is unable to remember all locations, he must concentrate on the most prominent among them lest he become confused and fail to effectively locate any of them at all.

(3) <u>Evaluating Aiming Points</u>. Targets which disappear behind good aiming points are easily remembered. Targets with poor aiming points are are easily lost. Assuming that two such targets are of equal value or are equally dangerous to the sniper, he should engage the poor aiming point first.

3. <u>Target Detection</u>. Depending on the type of enemy the sniper is employed against, the difficulty in locating that enemy will range from difficult in detecting a carefully moving patrol, to almost impossible in detecting scouts, or other snipers.

a. <u>Indicators</u>. In the class on camouflage, we learned how the sniper keeps from being detected. We must now see what indicators the sniper looks for to spot the enemy.

(1) <u>Movement</u>. With careful viewing of an area, movement is what will give away many targets to the observer. The techniques used in the hasty search will provide the best means for locating moving targets.

(2) <u>Improper Camouflage</u>. The improper use of camouflage usually provides indicators which reveal the majority of targets detected on the battlefield. However, many times an observation post or enemy firing position will blend almost perfectly with the natural background. Only the closest scutiny and observance of every object will reveal these positions. The three groups of improper camouflage to look for are shine, outline, and contrast.

(a) <u>Shine</u>. Shine may come from many sources, such as eye glasses, reflective metal, optical devices, pools of water, even the natural oils from the skin reflect light. Shine may last only a second, so the sniper has to be alert to observe it.

(b) <u>Outline</u>. Most enemy soldiers will camouflage themselves, their equipment and positions to break up their distinctive outlines, so the sniper, while observing, must be able to detect and identify an object by seeing only parts or bits of it, and from unusual angles.

(c) <u>Contrast</u>. An unusual color standing out against it's background may be a sign either of dying vegetation, used as camouflage, or of a piece of improper camouflage. A small patch of fresh soil may be an indicator of a camouflaged position. Communications wire that isn't buried is often easily seen, and it many times leads to an occupied position. While observing, anything that looks unusual or out of place should be studied in in minute detail. This kind of curiosity will greatly increase the chances of spotting hidden enemy.

(3) __Locating Hidden Shooters__. In the support of combat units, a
sniper may be called upon to locate and neutralize hidden riflemen or snipers.
One method to locate them is to use the sound of the bullet and discharge of
the rifle.

(a) __Crack and Thump__. When a rifle is fired from a distance toward
your direction, the first sound heard will be the crack of the bullet flying at
supersonic speed, as it passes by an object or passes overhead. More than one
crack may be heard if the bullet passes several objects. This crack is always
followed by a lower sounding thump, which is the discharge of the rifle.

(b) __Retracing the Shot__. At short range, the time between the
two sounds will be very short, and may even merge into one, but as the range
becomes greater, the time lapse between the crack and thump will increase.
With practice, one can determine the following:

(1) __Distance of the Firer__. The time lag between the crack
and thump can be quickly converted into the range at which the weapon is fired.
A one second lapse is about 600 yards, half a second is about 300 yards.

(2) __Determining Direction__. At first, it is natural to look
in the direction from which the crack of the bullet is heard, but it is necessary
to use the crack to time ones senses so that the direction of the following
thump can be determined.

(3) __Locating the Firer__. By watching in the direction of the
thump at the indicated range, the observer will usually see the smoke or flash
of a second or third shot fired.

(4) __Considerations__. If there were several shots being fired
from both sides, the sounds may interfere with one and other and the direction
finding will be very difficult. Also, sound will bounce off objects like
buildings or treelines to make direction finding confusing.

4. __Target Selection__. A sniper selects targets according to their value in a
given situation. Targets will have different importance if the sniper is in
support of infantry units than if he is working singularly. Certain enemy
personnel and equipment can be justifiably listed as key targets, but their
real worth must be decided by the sniper in relation to the circumstances in
which he locates them.

a. __Key Targets__.

(1) __Officers__. Officers of many countries can be distinguished by
plainly visible rank insignia. Officers may wear uniforms differing from those
of their troops in material, cut, color, and collar or cuff insignias. They
may carry binoculars, pistols, or map cases, and they will often have radiomen
near them. During a march, they may occupy random positions, or they may ride
in lead vehicles or staff cars.

(2) __Non-Commissioned Officers__. NCO's may be identified by their stripes
and their general actions in leading and directing troops. If troops are in a
formation, a higher ranking man will usually be doing the talking. In a bivuac
area, NCO's may have separate living, or eating areas, or will be at the end of
a chow line. In the Soviet-Warsaw-Pact Countries, NCO's may be identified by
the stripes or stars on their epaulets.

(3) <u>Scouts</u>. Scouts may be identified by the mode of their employment and usually by their working in small teams. They may be natives of the country, and familiar with the terrain. They work alone, in teams, or augmented with infantry troops. They are natural enemies of the sniper by the fact that they may stalk him on equal terms.

(4) <u>Crew-Served Weapons' Gunners</u>. They may be identified by their proximity to the weapon. The gunners to look for are machine gunners, mortar crews, recoilless rifle crews, and possibly artillery crews. Taking out these personnel will not only stop the weapon, but also do away with trained personnel. Sometimes, weapons crews will have a special insignia.

(5) <u>Tank Commanders</u>. They may be identifed by their position on the turret. The insignia in many Communist countries that is worn is simply, a tank.

(6) <u>Recruits</u>. Inexperienced troops may be identified by their lack of caution. If several of them are hit, others may panic and run. They may not camouflage themselves as well as others and may make a lot of careless moves.

(7) <u>Communications Personnel</u>. They may be identified by their equipment and employment. The sniper may lay a productive ambush by cutting wires and waiting for repairmen. If messengers are intercepted and shot, they should be searched if at all possible. The radio and radioman are prime targets. If the radioman is downed, attempts will be made to recover the radio, thus exposing more targets. The sniper should then destroy the radio when it is no longer needed. Soviet countries use a bolt of lightning as an insignia for communications.

(8) <u>Snipers</u>. They may be identified by the manner of their employment or their possession of telescopic sighted rifles, although many snipers are armed with iron sight rifles. Snipers are difficult to locate and are usually not seen until they have violated a "target indicator" principle or have fired their weapon. In searching for or firing on a sniper, one must be careful not to become a target himself. An enemy sniper is one of the few people who has been trained to react directly to another sniper.

(9) <u>Observers</u>. They may be identified by their use of optical equipment. They also may operate from a forward OP and direct artillery fire.

b. <u>Considerations of Selection</u>. The sniper's choice of targets may sometimes be forced upon him. He may lose a rapidly moving target if he waits to identify it in detail, and he must, of course, consider any enemy who threatens his position as a very high value target. When he is able to make free choice, however, the sniper will consider several factors:

(1) <u>Distance</u>. Even though he is capable of hitting a human at a range of 1,000 yards, the sniper should not risk such a distant shot without special reason. A normal shot would probably be between 600 and 1000 yards. 800 would be a good medium. Also, a sniper should never fire at a target under 300 yards because of the danger factor of being spotted and fired on.

(2) <u>Multiple Targets</u>. The sniper should carefully weigh the possible consequence of shooting at one of a number of targets, especially when he cannot identify them in detail. He may trade his life for an inconsequential target by putting himself in a position where he has to fire repeatedly in self-defense. Therefore, the sniper should never shoot more than three rounds from one position. The only time a sniper might risk more than three shots is if he is absolutely sure he is undetected, or if the target is of extremely high military or political rank.

(3) A well placed shot can disable crew-served weapons, radios, vehicles, and other equipment. There are many parts on equipment to fire at, such as tires or fuel tanks on vehicles, sights on artillery or recoilless rifles, mortar tubes, machine gun feed boxes and so on. Such equipment, however, may serve as bait and allow the sniper to make repeated kills of gunners or operators while keeping the equipment idle at the time. The equipment can be disabled at the sniper's later discretion.

(4) <u>Intelligence Collection</u>. Intelligence collection is an important collateral function of the sniper. When in location close to the enemy, he must be very judicious in his decision to open fire. He may interrupt a pattern of activity which, if observed longer, would allow him to report facts which would far outweigh the value of his kill. The kill could be made a short time later. The well-trained sniper will sensibly evaluate such situations.

<u>OPPORTUNITY FOR QUESTIONS AND COMMENTS</u>

<u>SUMMARY</u>

1. <u>Reemphasize</u>. During this period of instruction, we have discussed the initial steps leading up to target detection such as proper position selection, searching, and indexing targets. The sniping position we learned, if selected properly, is a primary source of the sniper's safety and an aid in accomplishing his mission. Indexing targets quickly and accurately is important in that targets may not appear for long and their position must be relayed between the observer and sniper.

In target detection, we learned the main indicators which will give away the enemies position to the sniper. They are movement or improper camouflage. Outline, shine, and contrast are three types of improper camouflage. Also in detecting, we covered finding a hidden shooter by listening to the crack and thump of the round and weapon, remembering a half second delay is about 300 yards and a second delay is about 600 yards.

Last, we covered target selection in which we learned the targets of greatest interest to the sniper. A sniper should always look to engage the key targets, but a given situation will decide what the key targets are.

2. <u>Remotivate</u>. We have seen how the sniper detects targets, then selects his victims. A sniper must be shrewd enough in "setting the stage" for the enemy to be where he can best be engaged to do him the greatest possible harm with no chance of reprisal against the sniper.

UNITED STATES MARINE CORPS
SCOUT/SNIPER INSTRUCTOR SCHOOL
Marksmanship Training Unit, Weapons Training Battalion
Marine Corps Development and Education Command
Quantico, Virginia 22134

SI0013
L, D, A

RANGE CARD, LOG BOOK AND
FIELD SKETCHING

(DATE)

DETAILED OUTLINE

INTRODUCTION

1. **Gain Attention**. The primary mission of a sniper is to deliver precision fire on selected targets from concealed positions. His secondary mission is to collect information about the enemy. To do this, he must primarily be observant, first to locate prospective targets and second to be able to identify what he sees. However observant he may be, the sniper cannot be expected to exercise the sheer feat of memory necessary to remember the ranges to all possible targets within his area of observation, or to recall all tidbits of information he may come across. The means designed to assist him in this task are the range card, the log book and the field sketch.

2. **Motivate**. Today you are going to learn how to record the data which you will need to accomplish both your primary and your secondary mission, thereby greatly enhancing your chance of acheiving a first round hit and of collecting useful and useable raw intelligence.

3. **State Purpose and Main Ideas**

   a. **Purpose**. To introduce you to the range card, log book and field sketch as used by the sniper in recording range estimates and in collecting information about the enemy.

   b. **Main Ideas**. To describe the preparation of range cards and their relationship to field sketches, and to teach how to draw adequate field sketches. Also to teach the components of various tactical reports for inclusion in the sniper log book.

4. **Learning Objectives**. Upon completion of this period of instruction, the student will be able to:

   a. Prepare a Range Card

   b. Prepare a Log Book

   c. Prepare a Field Sketch

<u>BODY</u>

1. The range card is a handy reference which the sniper uses to make rapid, accurate estimates of range to targets which he may locate in the course of his observations.

    a. (Show slide #1, Field Expedient Range Card) This slide illustrates a range card which a sniper might have prepared after his arrival at a point of observation. The card is drawn freehand and contains the following information:

        (1) Relative locations of dominating objects and terrain features.

        (2) Carefully estimated, or map measured, ranges to the objects or features.

        (3) The sniper's sight setting and holds for each range.

    b. (Show slide # 2, Prepared Range Card) Prior to departure on a mission, the sniper can prepare a better range card, shown here. Upon arrival in position, he draws in terrain features and dominant objects. To avoid preparing several cards for use in successive positions, the sniper can cover a single card with acetate and use a grease pencil to draw in the area features. Copies of the prepared range card should be prepared and used whenever possible.

    c. <u>Use of the Range Card</u>

        (1) <u>Holding</u>. (Show slide # 2, Prepared Range Card) The sniper locates a target in the doorway of the house at 10 o'clock from his position. From his card, he quickly determines a range of 450 yards and holds at crotch level. He centers the crosshairs on the crotch, fires, and hits the target in the center of the chest.

        (2) <u>Sight Setting</u>. The sniper locates a target on the roof of the house at one o'clock from his position. He notes the sight setting 61, applies that sight setting, and fires.

2. <u>The Field Sketch</u>.

    a. The field sketch is a drawn reproduction of a view obtained from any given point, and it is vital to the value of a sniper's log and range card, that he be able to produce such a sketch. As is the case for all drawings, artistic ability is an asset, but satisfactory sketches can be produced by anyone, regardless of artistic skill. Practice is, however, essential and the following principles must be observed:

        (1) Work from the whole to the part. Study the ground first carefully both by eye and with binoculars before attempting a drawing. Decide how much of the country is to be included in the sketch. Select the major features which will form the framework of the panorama.

        (2) Do not attempt to put too much detail into the drawing. Minor features should be omitted, unless they are of tactical importance, or are required to lead the eye to some adjacent feature of tactical importance.

Only practice will show how much detail should be included and how much left out.

      (3) Draw everything in perspective as far as possible.

  b. Perspective. The general principles of perspective are:

      (1) The farther away an object is in nature, the smaller it should appear in the drawing.

      (2) Parallel lines receding from the observer appear to converge; if prolonged, they will meet in a point called the "vanishing point." The vanishing point may always be assumed to be on the same plane as the parallel lines. Thus, railway lines on a perfectly horizontal, or flat, surface, receding from the observer will appear to meet at a point on the horizon. If the plane on which the railway lines lie is tilted either up or down, the vanishing point appears to be similarly raised or lowered. (Slide # 3) Thus, the edges of a road running uphill and away from the observer will appear to converge on a vanishing point above the horizon, and if running downhill, the vanishing point will appear to be below the horizon. (Slide # 4)

  c. Conventional Shapes. Roads and all natural objects, such as trees and hedges, should be shown by conventional outline, except where peculiarities of shape make them useful landmarks and suitable as reference points. This means that the tendency to draw actual shapes seen should be suppressed, and conventional shapes used, as they are easy to draw and convey the required impression. Buildings should normally be shown by conventional outline only, but actual shapes may be shown, when this is necessary to ensure recognition, or to emphasize a feature of a building which is of tactical importance. The filling in of outlines with shadowing, or hatching, should generally be avoided, but a light hatch may sometimes be used to distinguish wooded areas from fields. Lines must be firm and continuous.

  d. Equipment. The sniper should have with him the following items:

      (1) Suitable paper in a book with a stiff cover to give a reasonable drawing surface.

      (2) A pencil, preferably a No. 2 pencil with eraser

      (3) A knife or razor blade to sharpen the pencil

      (4) A protractor or ruler, and

      (5) A piece of string 15" long.

  e. Extent of Country to be Included. A convenient method of making a decision as to the extent of the country to be drawn in a sketch is to hold a protractor about 11 inches from the eyes, close one eye, and consider the section of the country thus covered by the protractor to be the area sketched. The extent of this area may be increased or decreased by moving the protractor nearer to, or farther away from, the eyes. Once the most satisfactory distance has been chosen, it must be kept constant by a piece of string attached to the protractor and held between the teeth.

f. <u>Framework and Scale</u>. The next step is to mark on the paper all outstanding points in the landscape in their correct relative positions. This is done by noting the horizontal distance of these points from the edge of the area to be drawn, and their vertical distance above the bottom line of this area, or below the horizon. If the horizontal length of the sketch is the same size as the horizontal length of the straight edge of the protractor, the horizontal distances in the picture may be gotten by lowering or raising the protractor and noting which graduations on its straight edge coincide with the feature to be plotted; the protractor can then be laid on the paper and the position of the feature marked against the graduation noted. The same can be done with vertical distances by turning the straight edge of the protractor to the vertical position.

g. <u>Scale</u>. The eye appears to exaggerate the vertical scale of what it sees, relative to the horizontal scale, i. e. things look taller than they are. It is preferable, therefore, in field sketching to use a larger scale for vertical distances than for horizontal, in order to preserve the aspect of things as they appear to the viewer. A suitable exaggeration of the vertical scale relative to the horizontal is 2:1, which means that every vertical measurement taken to fix the outstanding points in landscape should be doubled, while the horizontal measurements of the same points are plotted as read.

h. <u>Filling in the Detail</u>. When all the important features have been plotted on the paper in their correct relative positions, the intermediate detail is added, either by eye, or by further measurement from these plotted points. In this way, the sketch will be built up upon a framework. All the original lines should be drawn in lightly. When the work is completed, it must be examined carefully and compared with the landscape, to make sure that no detail of military significance has been omitted. The work may now be drawn in more firmly with darker lines, bearing in mind that the pencil lines should become darker and firmer as they approach the foreground.

i. <u>Conventioal Representation of Features</u>. The following methods of representing natural objects in a conventional manner should be borne in mind when making the sketch:

(1) <u>Prominent Features</u>. The actual shape of all prominent features which might readily be selected as reference points when describing targets, such as oddly shaped trees, outstanding building, towers, etc... should be shown if possible. They must be accentuated with an arrow and a line with a discription, e. g. Prominent tree with large withered branch.

(2) <u>Rivers</u>. Two lines diminishing in width as they recede should be used.

(3) <u>Trees.</u> These should be represented by outline only. Some attempt should be made to show characteristic shape of individual trees in the fore-ground.

(4) <u>Woods</u>. Woods in the distance should be shown by outline only. In the foreground, the tops of individual trees may be indicated. Woods my be shaded, the depth of shadowing becomes less with distance.

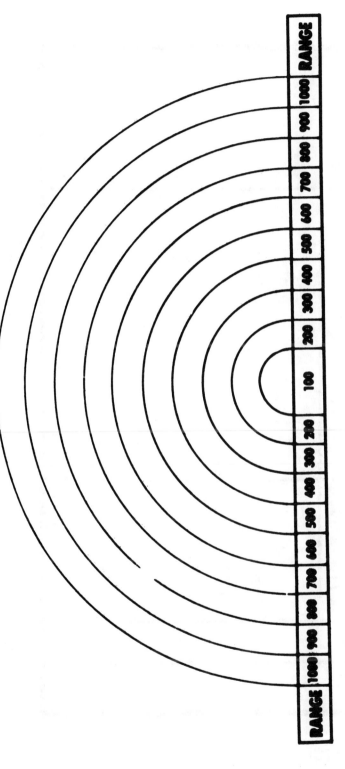

SNIPERS RANGE CARD

| RANGE | 1000 | 900 | 800 | 700 | 600 | 500 | 400 | 300 | 200 | 100 | 200 | 300 | 400 | 500 | 600 | 700 | 800 | 900 | 1000 | RANGE |

GRID COORDINATE OF POSTIONS _____

METHOD OF OBTAINING RANGE _____

MADE OUT BY _____

DATE _____

OBSERVATION LOG

Originator _____ Date _____ Tour of Duty _____ Sheet ____ of ____

Position _____ Visibility _____

Serial	Time	Grid Coordinate GR / Brg and Range	Event	Action or Remarks

(5) _Roads_. Roads should be shown by a double continuous line diminishing in width as it recedes.

(6) _Railways_. In the foreground, railways should be shown by a double line with small cross lines (which represent the ties) to distinguish them from roads; in the distance, they will be indicated by a single line with vertical ticks to represent the telegraph poles.

(7) _Churches_. Churches should be shown on outline only, but care should be taken to denote whether they have a tower or a spire.

(8) _Towns & Villages_. Definite rectangular shapes denote houses; towers, factory chimneys and prominent buildings should be indicated where they occur.

(9) _Cuts & Fills_. These may be shown by the usual topographic symbols, ticks diminishing in thickness from top to bottom, and with a firm line running along the top of the slope in the case of a cut.

(10) _Swamps and Marshland_. They may be shown by the conventional topographic symbols.

j. _Other Methods of Field Sketching_.

(1) _The Grid Window_. A simple device which can help a great deal in field sketching can be made by taking a piece of cardboard or hard plastic and cutting out of the center of it, a rectangle 6" x 2". A piece of clear plastic sheeting or celluloid is then pasted over the rectangle. A grid of squares of ½" size is drawn on the plastic sheeting. You now have a ruled plastic window through which the landscape can be viewed. The paper on which the drawing is to be made is ruled with a similar grid of squares. If the frame is held at a fixed distance from the eye by a piece of string held in the teeth, the detail seen can be transferred to the paper square by square.

(2) _The Compass Method_. Another method is to divide the paper into strips by drawing vertical lines denoting a fixed number of mils of arc and plotting the position of important features by taking compass bearings to them. This method is accurate but slow.

3. _The Sniper Log Book_. The sniper log is a factual, chronological record of his employment, which will be a permanent source of operational data. It will provide information to intelligence personnel, unit commanders, other snipers and the sniper himself. Highly developed powers of observation are essential to the sniper, as explained in earlier lessons. Because of this, he is an important source of intelligence whose reports may influence future operations, and upon which many lives may depend.

a. _Data to be Recorded_. The log will contain at a minimum, the following information:

(1) Names of observers,
(2) Hours of observation,
(3) Data and Position (Grid Coordinates)
(4) Visibility
(5) Numbered observations in chronological order,
(6) Time of observation,

(7) Grid reference of observation,

(8) Object seen, and

(9) Remarks or action taken.

b. Supplementary Materials. The sniper log is always used in conjunction
with the field sketch. In this way, not only does the sniper have a written
account of what he saw, but also a pictorial reference showing exactly where
he sighted or suspected enemy activity. If he is then relieved in place, the
new observer can more easily locate earlier sightings, by comparing the field
sketch to the landscape, than he could solely by use of grid coordinates.

OPPORTUNITY FOR QUESTIONS AND COMMENTS

SUMMARY

1. Reemphasize. During this period of instruction, we have looked at the Range
Card, The Field Sketch, and The Log Book. Each was presented as an entity in
itself and as how each related to the other.

2. Remotivate. The primary and secondary missions of the sniper were mentioned
at the beginning of this class. The primary mission can be fulfilled without
use of a log book and the secondary mission can be accomplished without recourse
to a range card. However, to be a complete Marine Scout/Sniper, you must be able
to prepare a useable field sketch and range card to ensure accurate shooting and
a thorough log book to collect all available intelligence data.

SCOUT / SNIPER
QUALIFICATION SCORE CARD

Name	Rank	SSN	Unit

Stage	Yd Line	Tgt Type	1	2	3	Tot Score
1	300	S				
2	300	M			✕	
3	500	S				
4	500	M			✕	
5	600	S				
6	600	M			✕	
7	700	S				
8	700	M			✕	
9	800	S				
10	800	M			✕	
					TOTAL	

PASS / FAIL
1st / 2nd

SCOUT / SNIPER

PIT QUALIFICATION SCORE CARD

Pit Puller Name	Rank	Tgt No

Stage	Yd Line	Tgt Type	Score		Tot Score	Int
1	300	S				
2	300	M		✕		
3	500	S				
4	500	M		✕		
5	600	S				
6	600	M		✕		
7	700	S				
8	700	M		✕		
9	800	S				
10	800	M		✕		
			TOTAL			

PASS	/	FAIL

1st	/	2nd

COURSE NO.	
UNIT & ADDRESS	

SSN	RANK	NAME

GRADING A = OUTSTANDING   B = VERY GOOD   C+ = ABOVE AVERAGE   C = AVERAGE   F = FAIL

WEEK

	1	2	3	4	5	6	REMARKS
INTEREST & ENTHUSIASM							
CAN HE ABSORB INSTRUCTION							
STANDARD OF KNOWLEDGE							
RANGE ESTIMATION							
MAP READING AND AIR PHOTOGRAPHY							
OBSERVATION							
CONCEALMENT							
STALKING							
SHOOTING							

WEEKLY REPORTS:

WEEK	COURSE OFFICER	SATIS PROG	REMARKS	(OTHER)
1				
2				
3				
4				
5				
6				

WEEK

	WEEK 1 MON	TUES	WED	THUR	FRI	WEEK 2 MON	TUES	WED	THUR	FRI	WEEK 3 MON	TUES	WED	THUR	FRI	WEEK 4 MON	TUES	WED	THUR	FRI	WEEK 5 MON	TUES	WED	THUR	FRI	WEEK 6 MON	TUES	WED	THUR	FRI
RANGE ESTIMATION																														
MAP READING & AIR PHOTOGRAPHY																														
OBSERVATION																														
CONCEALMENT																														
STALKING																														
SHOOTING																														

* ASSISTED DURING INSTRUCTION

TEST RESULTS

	RANGE ESTIMATION 1	2	MAP READING 1	2	OBSERVATION 1	2	CONCEALMENT 1	2	STALKING 1	2	SHOOTING 1	2	WRITTEN 1	2
HPS														
SCORE														
QUALIFICATION														
FINAL QUALIFICATION														

STATIONARY TARGETS

NAME
RANK
SSN
UNIT

DATE	YD LINES	1	2	3	4	5	6	7	8	9	10	11	12	13	14	15	16	17	18	19	20	21	22	23	24	25	TP	PP
	200																											
	500																											
	600																											
	300																											
	500																											
	600																											
	300																											
	500																											
	600																											
	300																											
	500																											
	600																											
	300																											
	500																											
	600																											
	300																											
	500																											
	600																											

COMMENTS

DATE	YD LINES	1	2	3	4	5	6	7	8	9	10	11	12	13	14	15	16	17	18	19	20	21	22	23	24	25	TP	PP
	700																											
	800																											
	900																											
	1000																											
	700																											
	800																											
	900																											
	1000																											
	700																											
	800																											
	900																											
	1000																											
	700																											
	800																											
	900																											
	1000																											

SCORING

THE VALUE OF EACH HIT WILL BE DETERMINED BY THE FORMULA V = R÷100

V = THE VALUE OF THE HIT,

R = THE RANGE

	R	SCORE
I.E.	V = 100÷100 =	1
	V = 200÷100 =	2
	V = 700÷100 =	7

A MISS WILL BE SCORED AS ZERO

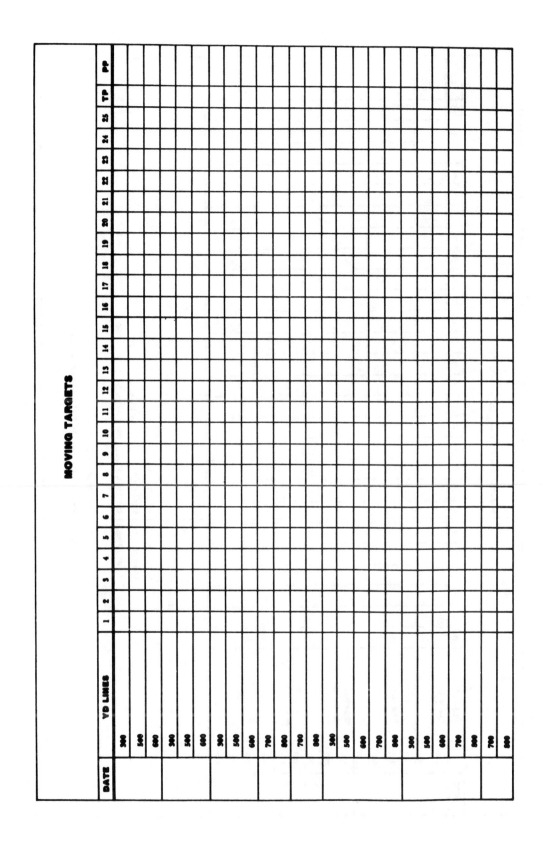

ISS AND ILLUMINATION FIRING

DATE	YD LINES		1	2	3	4	5	6	7	8	9	10	11	12	13	14	15	16	17	18	19	20	21	22	23	24	25	TP	PP
	1st ISS	100																											
		200																											
		300																											
	2nd ISS	100																											
		200																											
		300																											
	MOVING ILLUMINATION	200																											
		300																											
		500																											
	MOVING ILLUMINATION	200																											
		300																											
		500																											
	STATIONARY ILLUMINATION	200																											
		300																											
		500																											

SCORING

ISS 5 POINTS FOR 1st RD HIT

MOVING ILLUMINATION: FORMULA V— R+100

STATIONARY ILLUMINATION: FORMULA V — R+100

MISS WILL BE SCORED AS ZERO

PASSING SCORE 80% OF THE TOTAL POINTS AVAILABLE

SNIPER INSTRUCTION COURSE NO._____
RANGE ESTIMATION EXCERCISE NO._____
STUDENT SCORE SHEET

RANK & NAME_____

PROBLEM NO. 1_____**YARDS**

PROBLEM NO. 2_____**YARDS**

PROBLEM NO. 3_____**YARDS**

PROBLEM NO. 4_____**YARDS**

PROBLEM NO. 5_____**YARDS**

PROBLEM NO. 6_____**YARDS**

PROBLEM NO. 7_____**YARDS**

PROBLEM NO. 8_____**YARDS**

DISTANCES JUDGED ACCEPTEDLY_____

PASS or FAIL

NAME_____ KIM GAME NO._____

NO.	ITEM IN DETAIL	SCORE
1		
2		
3		
4		
5		
6		
7		
8		
9		
10		
11		
12		
13		
14		
15		
16		
17		
18		
19		
20		
	TOTAL	

<u>TRANSITION</u>. The sniper must not only estimate his target range, but also it's speed and angle of travel relative to his line of sight in order to determine the correct lead. (TA # 1 & 2)

 (1) <u>Angle of Target Movement</u>. TA # 1 shows the leads required for a target moving at a 90-degree angle to the sniper. In most cases, however, targets will be moving at some other angle toward or away from the sniper's position. A method for estimating the angle of movement is as follows:

  (a) <u>Full Lead Target</u>. (TA #2) When the target is moving across the observer's front and only one arm and one side are visible, the target is moving at or near an angle of 90 degrees and a full value lead is necessary.

  (b) <u>Half Lead Target</u>. When one arm and two-thirds of the front or back are visible, the target is moving at approximately a 45 degree angle and a one-half value lead is necessary.

  (c) <u>No Lead Target</u>. When both arms and the entire front or back are visible, the target is moving directly toward or away from the sniper and will require no lead.

 (2) <u>Double Leads</u>. The leads previously mentioned hold true for a right-handed shooter firing on a target moving from his right to left. If the target is moving from left to right, the lead must be doubled due to a natural hesitation in followthrough when swinging against the shooting shoulder. This hesitation is extremely difficult to overcome even by the most experienced shooters.

<u>OPPORTUNITY FOR QUESTIONS</u>

<u>SUMMARY</u>

1. <u>Reemphasize</u>. During this period of instruction, we have discussed the two different methods most often used to lead a moving target and emphasized that it was important to stick with one method and not fluctuate back and forth between the two.
 We covered the required leads that should be used to hit a moving target out to 800 yards.
 In conclusion, we discussed how to estimate angle of target movement and use of a full lead and half lead. Double leads were covered and the situation was covered as when to apply them.

2. <u>Remotivate</u>. As you can see, the sniper must now become proficient in his ability to judge distance, how fast his target is moving, and at what angle the target is moving with respect to him and still put that first round on target at ranges out to 800 yards.

UNITED STATES MARINE CORPS
SCOUT/SNIPER INSTRUCTOR SCHOOL
Marksmanship Training Unit, Weapons Training Battalion
Marine Corps Development and Education Command
Quantico, Virginia 22134

SI 0014
L, D, A

_____
(DATE)

LEADS

DETAILED OUTLINE

INTRODUCTION

1. **Gain Attention**. You and your partner have been in position for several days without any luck at all, and are just packing it in when your partner catches sight of someone moving down a dry river bed, approximately 675 to 700 yards down range. You both decide that he is moving at about a 45 degree angle to you, at an average pace. You obtain what you think is the proper hold and lead for that range and squeeze the shot off. Your partner doesn't say anything, but looks at you and winks.

2. **Motivate**. First round kill is the name of the game. Being snipers, you could very well be placed in this situation and when you are, will be expected to put that round right where it belongs on a moving target out to 800 yards.

3. **State Purpose and Main Ideas**.

   a. **Purpose**. The purpose of this period of instruction is to provide the student with the knowledge of the proper leads to be used to hit a moving target (walking and running) at ranges from 100 to 800 yards.

   b. **Main Ideas**. The main ideas to be discussed are the following:

      (1) Methods of Leading a Moving Target
      (2) Angle of Target Movement
      (3) Normal Leads
      (4) Double Leads

4. **Learning Objectives**. Upon completion of this period of instruction, the student will, without the aid of references, understand and be able to demonstrate the proper lead necessary to hit a moving target at ranges from 100 to 800 yards.

TRANSITION. The best example of a lead can be demonstrated by a quarterback throwing a pass to his reciever. He has to throw the ball at some point down field to where the reciever has not yet reached. The same principle applies in shooting at a moving target with the sniper rifle.

BODY

1. **LEADS**. Moving targets are the most difficult to hit. When engaging a target which is moving laterally across his line of sight, the sniper must concentrate on moving his weapon with the target while aiming at a point some distance ahead. Holding this "lead", the sniper fires and follows through with the movement after the shot. Using this method, the sniper reduces the possibility of missing, should the enemy suddenly stop, hit the deck, or change direction. The following is a list of ranges and leads used to hit moving targets both walking and running;

	WALKING	RUNNING
RANGE	LEAD	LEAD
100	Front edge of body	½ foot/body width
200	½ foot/body width	1 foot/body width
300	1 foot/body width	2 feet/body width
400	1½ feet/body width	3 feet/body width
500	2 feet/body width	4 feet/body width
600	2½ feet/body width	5 feet/body width
700	3 feet/body width	6 feet/body width
800	3½ feet/body width	7 feet/body width

Another method of leading a target, and one which is used extensively by the British, is known as the "point" lead. By "point lead", we mean the sniper selects a point some distance in front of his target and holds the crosshairs on that point. As the target moves across the horizontal cross-hair, it will eventually reach a point which is the proper lead distance from the center. At that instance, the sniper must fire his shot. This is a very simple method of hitting a moving target, but a few basic marksmanship skills must not be forgotten:

    a. The sniper must continue to concentrate on the crosshairs and not on the target.

    b. The sniper must continue to squeeze the trigger and not jerk or flinch prior to the shot being fired.

    c. Some snipers tend to start with this method, but begin to track the target once it reaches that magic distance and then fire the shot. Use one of the two methods and stick with the one which you are confident will get that shot on target. (The instructor should draw these methods of leading on the chalkboard to better illustrate.)

UNITED STATES MARINE CORPS
SCOUT/SNIPER INSTRUCTOR SCHOOL
Marksmanship Training Unit, Weapons Training Battalion
Marine Corps Development and Education Command
Quantico, Virginia 22134

SI0015
L, D

_____
(DATE)

OCCUPATION
AND
SELECTION OF POSITIONS
TITLE

DETAILED OUTLINE

INTRODUCTION

1. **Gain Attention.** Relate story of Russian super-sniper Vassili Zaitsev and German super-sniper Major Konigs at the Battle of Stalingrad. (Excerpts from Enemy at the Gates by William Craig.)

2. **Motivate.** Simply stated, the snipers mission is to see without being seen and to kill without being killed.

3. **State Purpose and Main Ideas.**

   a. **Purpose.** The purpose of this period of instruction is to provide the student with the knowledge required to select and occupy a position.

   b. **Main Ideas.** The main ideas to be discussed are the following:

       (1) Position Selection
       (2) Hasty Positions
       (3) Position Safety
       (4) Actions in Position

4. **Learning Objectives.** Upon completion of this period of instruction the student will:

   a. Identify those features which contribute to the selection of a position. i. e. cover, concealment, fields of fire, avenues of approach and withdrawal, etc.

   b. Determine, using maps, aerial photos and/or visual reconnaissance, the location of a suitable sniper position.

<u>TRANSITION</u>. To effectively accomplish their mission of supporting combat operations by delivering precision fire on selected targets the sniper team must select a position from which to observe and fire.

<u>BODY</u>

1. <u>Position Selection</u>. The sniper, having decided upon an area of operation, must choose a specific spot from which to operate. The sniper must not forget that a position which appears to him as an obvious and ideal location for a sniper will also appear as such to the enemy. He should avoid the obvious positions and stay away from prominent, readily identifiable objects and terrain features. (TA) The best position represents an optimum balance between two considerations.

    a. It provides maximum fields of observation and fire to the sniper.

    b. It provides maximum concealment from enemy observation.

2. <u>Hasty Positions</u>. Due to the limited nature of most sniper missions and the requirement to stalk and kill, the sniper team will in most cases utilize a hasty post. Considering the fundamentals of camouflage and concealment the team can acquire a hasty sniper post in any terrain. (TA) The principle involved when assuming a hasty position is to utilize a maximum of the team's ability to blend with the background or terrain and utilize shadows at all times. Utilizing the proper camouflage techniques, while selecting the proper position from which to observe and shoot, the sniper can effectively preclude detection by the enemy. (TA) While hasty positions in open areas are the least desirable, mission accomplishment may require assuming a post in an undesirable area. Under these circumstances, extreme care must be taken to utilize the terrain (ditches, depressions, and bushes) to provide maximum concealment. The utilization of camouflage nets and covers can provide additional concealment to avoid detection. There should be no limitation to ingenuity of the sniper team in selection of a hasty sniper post. Under certain circumstances it may be necessary to fire from trees, rooftops, steeples, under logs, from tunnels, in deep shadows, and from buildings, swamps, woods and an unlimited variety of open areas.

3. <u>Position Safety</u>. Selection of a well covered or concealed position is not a guarantee of the sniper's safety. He must remain alert to the danger of self-betrayal and must not violate the following security precautions.

    a. When the situation permits, select and construct a sniper position from which to observe and shoot. The slightest movement is the only requirement for detection, therefore even during the hours of darkness caution must be exercised as the enemy may employ night vision equipment and sound travels great distances at night.

b. The sniper should not be located against a contrasting background or near prominent terrain features, these are usually under observation or used as registration points.

c. In selecting a position, consider those areas that are least likely to be occupied by the enemy.

d. The position must be located within effective range of the expected targets and must afford a clear field of fire.

e. Construct or employ alternate positions where necessary to effectively cover an area.

f. Assume at all times that the sniper position is under enemy observation. Therefore while moving into position the sniper team should take full advantage of all available cover and concealment and practical individual camouflage discipline. i. e. face and exposed skin areas camouflaged with appropriate material. The face veil should be completely covering the face and upon moving into position the veil should cover the bolt receiver and entire length of the scope.

g. Avoid making sound.

h. Avoid unnecessary movement unless concealed from observation.

i. Avoid observing over a skyline or the top of cover or concealment which has an even outline or contrasting background.

j. Avoid using the binoculars or telescope where light may reflect from lenses.

k. Avoid moving foilage concealing the position when observing.

l. Observe around a tree from a position near the ground.

m. Stay in the shadow when observing from a sniper post within a building.

n. Careful consideration must be given to the route into or out of the post. A worn path can easily be detected. The route should be concealed and if possible a covered route acquired.

o. When possible, choose a position so that a terrain obstacle lies between it and the target and/or known or suspected enemy location.

p. While on the move and subsequently while moving into or out of position all weapons will be loaded with a round in the chamber and the weapon on safe.

4. <u>Actions in Position</u>. After arriving in position and conducting their hasty then detailed searches, the sniper team organizes any and all equipment in a convenient manner so it is readily accessable if needed. The sniper team continues to observe and collect any and all pertinent information for intelligence purposes. They establish their own system for observation, eating, sleeping, resting and making head calls when necessary. This is usually done in time increments of 30 to 60 minutes and worked alternately between the two snipers for the entire time they are in position, allowing one of the individuals to relax to some degree for short periods. Therefore it is possible for the snipers to remain effective for longer periods of time.

The sniper team must practice noise discipline at all times while occupying their position. Therefore arm and hand signals are widely used as a means of communicating. The following are recommended for use when noise discipline is of the utmost importance.

a. Pointing at oneself; meaning I, me, mine.

b. Pointing at partner; meaning you, your, yours.

c. Thumbs up; meaning affirmative, yes, go.

d. Thumbs down; meaning negative, no, no go.

e. Hands over eyes; meaning cannot see.

f. Pointing at eyes; meaning look, see, observe.

g. Slashing stroke across throat; meaning dead, kill.

h. Hands cupped together; meaning together.

i. Hand cupped around ear, palm facing forward; meaning listen, hear.

j. Fist; meaning stop, halt, hold up.

k. Make pumping action with arm; meaning double time.

OPPORTUNITY FOR QUESTIONS

SUMMARY

1. Reemphasize. During this period of instruction we discussed position selection and the two factors necessary to all positions (1) Provides maximum fields of observation and fire to the sniper. (2) It provides maximum concealment from enemy observation.

We then covered selection of hasty positions and that mission accomplishment might require assuming a position in an undesireable area. All available terrain should be used to provide maximum concealment under these circumstances.

In conclusion we covered a number of safety precautions to be considered while on the move and in the process of moving into and out of position.

2. Remotivate. How well the sniper team accomplishes the mission depends, to a large degree, on their knowledge, understanding, and application of the various field techniques or skills that allow them to move, hide, observe, and detect. These skills are a measure of the sniper's ability to survive.

UNITED STATES MARINE CORPS
SCOUT/SNIPER INSTRUCTOR SCHOOL
Marksmanship Training Unit, Weapons Training Battalion
Marine Corps Development and Education Command
Quantico, Virginia 22134

SI0016
L, D, A

_____
(DATE)

OFFENSIVE EMPLOYMENT

DETAILED OUTLINE

INTRODUCTION

1. **Gain Attention.** When the bombs have stopped falling, when the artillery grows silent, when the thundering roar of naval gunfire is a memory, the ground will still have to be assaulted and secured by the rifleman. When this time comes, the grunt is almost entirely on his own. Indirect fire weapons can not help in these last few murderous yards when a round, slightly short of its target, could kill the wrong men. And yet, the rifleman is not entirely without help. Supporting him is one of his own, a fellow infantryman, specially trained who will, with deadly accurate fire, keep the enemy's head down or kill him.

2. **Motivate.** If you complete this course successfully, you will be that final supporting arm. Naturally, your usefulness extends far beyond the final assault. You can expect to be very busy in contact remote, in consolidation and exploitation, and even on special assignment. However, you will still be the man with the rifle, upon whom that grunt depends more than even he realizes.

3. **State Purpose and Main Ideas**

   a. **Purpose.** To provide the student with knowledge of all phases of offensive combat and the employment of snipers in all of these phases.

   b. **Main Ideas.** To explain the snipers role in:

      (1) Movement to Contact
      (2) The Assembly Area
      (3) The Various Forms of Maneuver

4. **Learning Objectives.** Upon completion of this period of instruction, the student will be able to:

   a. Employ a battalion level sniper section in offensive operations.

   b. Advise supported unit commanders on all phases of sniper employment during offensive operations.

BODY

1. Sniper's Role.

    a. General. Infantry units, acting independently or as parts of larger forces, conduct offensive movements to contact, close with, and destroy the enemy. Snipers provide the infantry commander with an additional means of accomplishing his mission. They are capable of detecting and shooting long-range targets which could otherwise impede the progress of the defense.

    b. Tasks Common to All Offensive Operations. The tasks of snipers in the offensive role include:

        (1) Supporting the infantry by delivering accurate, long-range fire at:

            (a) Enemy automatic weapons emplacements or embrasures.

            (b) Enemy artillery forward observers.

            (c) Enemy personnel.

            (d) Enemy optical devices used for observation purposes.

            (e) Fleeing enemy personnel during the consolidation and exploitation phase.

        (2) Protecting the flanks of attacking units.

        (3) Covering by fire, gaps between attacking elements.

        (4) Participating in repelling counterattacks.

2. Offensive Combat. The ultimate purpose of offensive action is the destruction of the enemy's armed forces, imposition of the commander's will on the enemy, or the seizure of key terrain necessary to further accomplishment of the overall mission which becomes increasingly more important as contact becomes imminent.

3. Movement to Contact. Movement to contact is a tactical movement to gain contact with the enemy. The intent may be to establish initial contact with him or reestablish contact which has been lost. The movement to contact is terminated when physical contact with the enemy is gained or when the march units cross the line of departure.

    a. Movement. Snipers may conduct the movement to contact attached to the infantry battalion headquarters or attached to a company. The movement may be a covered or uncovered movement.

        (1) An uncovered movement is made by the leading elements of a force with the mission of gaining contact. Information of the enemy may or may not be available from friendly ground units to the front. Reconnaissance by the advancing units must be intensified to compensate for a lack of security elements provided by other forces. The uncovered movement ends when contact is gained or when information concerning the enemy warrants launching an attack. During an uncovered movement, sniper teams may be assigned anywhere that their observation capability can be used. The following assignments

are considered appropriate during an uncovered movement:

        (a) As members of a reconnaissance team on critical terrain features along the axis of advance.

        (b) As security in areas where there is probability of ambush.

    (2) A covered movement to contact is made when adequate security is provided by other forces. It normally ends when an assigned location is occupied and it is usually an administrative move.

    b. <u>Sequence of Movement</u>. The movement to contact is made in route column, tactical column, or approach march.

    (1) <u>Route Column</u>. Where enemy contact is remote, the movement is made administratively in route column. Units need not be grouped tactically and may move by various means over different routes. No specific sniper tasks are assigned during this phase. The route column regroups tactically when the commander's estimate of the probablilty of contact changes from remote to improbable. Maximum emphasis is placed on ease of control and speed of movement.

    (2) <u>Tactical Column</u>. When the probability of contact with the enemy changes from contact remote to contact improbable, units within the column are tactically grouped and adequate security to the front, flanks, and rear of the column is provided. The following units constitute security for the main body:

        (a) <u>Advance Guard</u>. The advance guard precedes the main advance. It protects against observation and surprise by the enemy. The advance guard is subdivided into a point, an advance party, and a support. Flank security is assigned by both the advance party and the support. Snipers normally accompany the advance guard. They may be with the support unit or any of the subordinate security elements.

        (b) <u>Advance Party</u>. The advance party is deployed forward by the advance guard and constitutes the reconnoitering element for the support. It provides its own point and flank security and is capable of coping with minor resistance. Snipers augment the observation capability of the advance party and also provide long-range fire at targets of opportunity. The mobility required of security elements and the terrain often precludes employment of other supporting arms. Sniper teams might very well constitute the only supporting arm available to engage long-range targets.

        (c) <u>Point</u>. The point is a small detachment sent forward by the advance party to give warning to enemy activity. Sniper teams may accompany the point to provide additional observation and firepower. Snipers should always augment a normal size point element; they should never be substituted for riflemen.

        (d) <u>Rear guard</u>. The rear guard consists of a rear point and the rear party. The rear guard has no support and is large enough to protect the rear of the tactical column. Assignment of sniper teams to the rear guard is dependent upon the terrain, probability of enemy contact, and the priority of need.

(e) Flank Guards

       1 Each major subdivision of the march column establishes flank security. Flank guards move abreast of their subdivision and parallel to it. They occupy successive key terrain features which cover approaches to the line of march. The flank guard prevents the enemy from bringing effective fire or observation to bear on the main body and will engage the enemy as necessary.

       2 The flank guard regulates its movement to the main body rate of march and must travel rapidly over great distances in shorter periods of time than the march column. Often the terrain over which they move is more difficult than that of the march route taken by the column. Frequent relief of the sniper teams attached to the flank guard is necessary. The relieving sniper team stations itself ahead of the flank guard, covers the approach to the flank guard, and relief is effected as the flank guard passes through. Sniper teams support by fire and observation movement between successive intermediate objectives and key terrain features dominating likely avenues of approach into the flank.

      (3) Approach March. As the commander's estimate of the probability of contact with the enemy changes from contact improbable to contact imminent, the march column increases its readiness for combat. Units in the column are task organized and tactically grouped for immediate deployment from the march column. Tactical considerations in the organization of the column outweigh other considerations. Sniper teams attached to the advance guard may be withdrawn and integrated into the tactical organization. The decision will depend upon the situation in general and, more specifically, the determination of which assignment has priority in terms of anticipated need. The sniper squad or platoon leader constantly evaluates the situation and is responsible for planning and advising the commander on employment of the sniper teams.

4. Assembly Area. The covered movement to contact normally terminates in the occupation of an assembly area where final preparation is made for the offense.

    a. Organization. Within the assembly area, elements of the unit are dispersed to reduce vulnerability to enemy fires. The unit takes advantage of all available cover and concealment, and an all-around defense is established. Sniper teams are positioned for observation into those areas which are most likely to be occupied by enemy snipers.

    b. Preparation. In the assembly area, final preparations for combat are made. The sniper leader must ensure timely rotation of sniper teams conducting the defense to provide all hands with the opportunity to rest and prepare. Preparations include the following:

      (1) Final coordination with the commander of the unit to which attached to clarify or amplify the sniper mission.

      (2) Thorough cleaning and checking of all gear, particularly the telescope, telescope mount, and ammunition.

      (3) Drawing of rations and water.

(4) Application of camouflage.

(5) Obtaining of radio call signs and frequencies.

(6) Participation in rehearsals or training.

5. Forms of Maneuver

a. **General.** There are four basic forms of offensive maneuver: frontal attack, envelopment, penetration, and the turning movement. The characteristics of the area of operations, the situation of the enemy, terrain, suitable avenues of approach, obstacles, and the enemy defensive posture influence the choice of maneuver.

b. **Sniper Assignment.** Sniper teams may be assigned to the rifle company on a daily basis or for the duration of a specific operation. The company commander will assign their specific tasks. Sniper teams may also be assigned to battalion headquarters, under the operational control of the battalion commander.

6. Frontal Attack. The frontal attack is designed to achieve tactical success along an entire front. The purpose of the attack is to exert pressure along the enemy front and push him off the objective with superior combat power.

a. **Sniper Team Employment.** The basic offensive mission of a sniper team in a frontal attack is to support the advance of the attacking infantry by precision fire. This is accomplished by:

(1) Having moved into position on the night before the assault, direct fire at any enemy key personnel prior to the start of the assault, thereby giving the enemy no time to replace them.

(2) Directing fire from carefully concealed positions at exposed enemy troops.

(3) Delivering fire into embrasures in enemy bunkers.

(4) Destroying enemy crew-served weapons and crews.

(5) Delivering long-range fire at targets located beyond the objective but directly opposing the advance. (Firing is normally shifted to deeper targets when the maneuvering infantry masks the fire.)

(6) Providing flank protective fire at targets threatening an exposed flank or at small, isolated resistance pockets which have been bypassed.

(7) Assisting in consolidation and exploitation by firing at targets threatening a counterattack or firing at fleeing enemy personnel.

(8) Following in trace of the assault force to eliminate any enemy snipers who may have concealed themselves along our avenues of approach to harass the rear of our forces.

b. _Priority of Targets_. Snipers should primarily be assigned specific point targets. The sniper team cannot engage all known or suspected enemy positions. In such cases, it will be necessary to successively neutralize enemy weapons or personnel having the greatest effect on accomplishment of the mission.

c. _Requirements for Firing Positions_. During the attack, the sniper team's primary position should be one which enables the snipers to cover the entire front of the objective and as much of the rear of the objective as possible. An alternate position should be preselected for occupation in case enemy fire makes the primary position untenable. Supplementary positions are required when it is anticipated that snipers may be required to engage other targets in addition to or in lieu of the primary targets. All three positions should offer as many of the following characteristics as possible:

(1) Clear fields of fire
(2) Cover and concealment
(3) Observation of as much of the objective area as possible.
(4) Sufficient height to preclude premature masking of fire by advancing friendly troops.

d. _Displacement_

(1) When the assault begins and fires are masked, the sniper team displaces as rapidly as possible to new positions to continue support of the attack.

(2) If the tactical situation requires uninterrupted fire and targets are still evident to the rear or flank of the objective, the sniper team will displace by echelon. One sniper will continue to provide supporting fire while the other moves forward to a newly selected position. Upon assumption of the new position, the other sniper will rejoin the team.

7. _Envelopment_. In the envelopment, the main attack is directed against the enemy flanks or immediate rear while a supporting attack exerts pressure along his front to fix him in position. There are two types of envelopment: the single envelopment and the double envelopment. The concept of sniper team employment is the same for both.

a. _Sniper Team Employment_. The basic offensive mission of a sniper team in an envelopment is to support the enveloping unit or units by precision fire. This is accomplished by:

(1) Directing precision fire from carefully concealed positions in the vicinity of the unit or units conducting the supporting attacks.

(2) Preventing the enemy from physically reinforcing the area in which the main attack is being conducted by delivering precision fire at enemy reinforcements.

(3) Engaging the targets as outlined in paragraph 6a.

b. Requirements for Firing Positions. During the envelopment, the best position from which to support the main attack is in the vicinity of the supporting unit. Although it is recognized that the greatest concentration of fire support is normally allocated to the main attack, the potential of the sniper rifle is best exploited in the support role. Attachment of the sniper team to rapidly moving assault units would not provide adequate time for establishing good firing positions from which to support the attack. The ideal firing position permits full fire to the front to deceive the enemy as to the true location of the main attack and also provides a field of fire into the area of the main attack.

c. Firing. As the enveloping unit is maneuvering, snipers deliver precision fire at exposed targets to the direct front. As the enveloping unit nears the final coordination line, sniper fires are shifted to the area of the main attack. As the main attack nears the assault position, supporting fires are normally ceased or shifted to permit the momentum of the assault to carry it through the objective. Snipers, however, continue to provide precision fire at targets which appear upon cessation of the supporting fires. Extreme caution is exercised to prevent possible ricochets and injury to fiendly personnel. Fires are shifted as the main attack starts through the objective.

d. Displacement. As the main attack secures the objective, sniper teams are displaced forward. They are positioned within the defensive posture to pursue the enemy by fire and assist in repelling counterattacks.

8. Penetration. A penetration is an attack through some portion of the enemy's position and is directed against an objective to his rear. It is chatacterized by an initial attack on a wide front to fix the enemy and deceive him as to the location of the main attack. The main attack is a powerful violent attack in considerable depth, launched on a narrow front.

a. Sniper Team Employment. The concept for employment of sniper teams during a penetration is similar to that for an envelopment. Snipers contribute to the accomplishment of the mission by directing the bulk of their fires into targets opposing the main attack.

b. Requirements for Firing Positions. The penetration is initially supported from sniper positions, located near the area of the intended breakthrough, which provides the essential alements of a good position as outlined in subparagraph 6c.

c. Firing. Prior to the attack, snipers engage targets directly opposing the main attack. Particular emphasis is placed on enemy observers and embrasures in bunkers which threaten the attack. As the attack commences, exposed enemy personnel are taken under fire. When the breakthrough occurs, fires are shifted to targets falling back as the penetration is widened.

d. Displacement. After the penetration is effected, sniper teams displace forward to support exploitation of the area immediately adjacent to the penetration. They are then in position to pursue the enemy by fire or assume firing positions to support an attack upon objectives to the rear of the enemy lines.

9. Turning Movement

   a. General. In the turning movement, the main attack passes around the main enemy force and seeks to secure an objective deep to the rear. Coincident with the maneuver of the main attack, a supporting attack exerts pressure on the front to divide his attention. The purpose of the turning movement is to compel the enemy to abandon his positions or divert major forces to meet a new threat to his rear.

   b. Sniper Team Employment. The wide separation between the supporting unit and the unit conducting the main attack necessitates attachment of sniper teams to each element. If sufficient sniper teams are not available, the priority of attachment should be to the unit conducting the main attack. The considerations governing firing positions, firing and displacement are similar in all respects to those exercised in other forms of maneuver.

10. Tank-Infantry Operations. Tanks are employed with infantry in a balanced tank-infantry team to exploit the mobility, firepower, speed, and shock action of the tank. Sniper teams provide long-range protection for the tanks from concealed enemy antitank weapons and their crews and tank-killer teams. Further assistance is rendered the tank element by observing and detecting tank targets and marking them with tracer rounds.

   a. Employment Limitations. The speed of tanks requires sniper teams to make frequent displacement forward in order to remain within support range. Terrain posses an additional limitation upon the effectiveness of sniper support. As the tanks move forward, fires are frequently masked by hills or other prominent terrain or manmade features.

   b. Sniper Team Employment. Snipers are effectively employed in the support of tank-infantry operations only when desirable employment conditions exist. Desirable employment conditions are defined as "any tactical situation in which the terrain does not mask or otherwise restrict the effective application of fire." Targets which can be effectively engaged by the sniper team are taken under fire and destroyed. Those targets which are invulnerable to small arms fire are destroyed by tank fire. Snipers mark targets at ranges up to 900 meters by tracer rounds. Active sniper team employment commences with the movement from the assembly area and continues through the consolidation phase.

   c. Methods of Employment. The infantry commander uses any combination of three three methods of attack in employing tanks. The three methods are: same axis, converging axes, and support by fire.

      (1) Same Axis. When maneuver, visibility, and fields of fire are restricted, tanks and infantry usually advance together in mutual support. Sniper teams occupy positions along the axis of advance which permit maximum observation and field of fire into the area immediately to the front and flanks of the tanks. The telescope, adjusted to high magnification, will readily detect the presence of enemy anti-tank positions and tank-killer teams which may be invisible to the tank-infantry team. Sniper observation is concentrated on terrain just beyond that which can be easily seen by the advancing tank-infantry team. This technique of tank-infantry employment permits close coordination and maximum mutual support between tank and infantry but sacrifices the speed and mobility of the tanks.

Consequently, sniper teams need not displace forward as often as when the tanks are alone. Sniper displacement should be by echelon to preselected locations to provide continuous support. As the tank-infantry team nears the objective area, increased enemy resistance becomes likely. Consequently, sniper teams must plan to be in a firing position from which they can effectively support the final stages of the assault.

(2) <u>Converging Axes</u>. Separate axes are used by tanks and infantry to approach the objective. Normally, the tanks will follow the terrain most appropriate for their employment while the infantry follows a route offering cover and concealment. If the terrain permits, snipers are attached in direct support of the tank movement augmenting a minimum size force of infantry. Rapidly moving tanks will normally preclude the assumption of stationary firing positions. The sniper team will follow in trace of the tanks, taking advantage of every opportunity to stop, observe, and detect concealed targets for the tanks. During the assault, the sniper team supports the operation by precision fire from positions which provide maximum observation and fields of fire. Upon occupation of the objective, they participate in the consolidation phase.

(3) <u>Support by Fire</u>. The support by fire method consists of an infantry attack which is supported by tank fire from stationary positions. It is considered the least desirable method of attack and is used only when conditions preclude assault by tanks. If the distance from the supporting tank position to the objective is within the effective range of the sniper rifle, snipers will normally be attached to the tank element to provide protection and additional observation for the tanks, mark targets with tracer rounds, and to deliver precision fire at targets of opportunity. If the tanks are beyond the effective range of the sniper rifle, the commander may leave the sniper team in support of the tanks, place them in a position between the tanks and the objective, or attach them to the infantry in a conventional role.

d. <u>Plan of Attack</u>. The plan of attack embodies a scheme of maneuver and a fire support plan, developed concurrently, based on an estimate of the situation

(1) <u>Scheme of Maneuver</u>. A sniper team representative should be present to conduct a joint reconnaissance with the infantry and tank commander if the situation permits. The sniper representative makes appropriate sniper employment recommendations based upon the indicated scheme of maneuver.

(2) <u>Fire Support Plan</u>. The sniper teams' unique capabilities play a significant and prominent role in the fire support plan of a tank-infantry team. Fires must be planned to protect the tanks from tank-killer teams and known or suspected antitank positions. Fires to protect and cover tank movement are particularly important when the tank must traverse over terrain not previously uncovered by the infantry. The observation and precision fire capability of the sniper team lend invaluable assistance to the commander in formulating a tank-infantry fire support plan.

11. Mechanized-Motorized Attack

a. __General__. The infantry unit is mechanized when it is supported by amphibious vehicles and tanks for the purpose of conducting land combat. Mechanized infantry operations are characterized by rapid mounted movement to an area forward of the line of departure for subsequent continuation of the attack on foot. Movement is made by the infantry mounted in amphibious tractors and supporting tanks using the same or multiple axes. When movement is by multiple axes, priority of sniper attachment is to the amphibious tractor element. Mechanized infantry attacks are organized to accomplish one or more of the following missions:

   (1) Rapid seizure of deep objectives.
   (2) Envelopment and seizure of enemy positions.
   (3) Pursuit and/or cutoff of withdrawing enemy.
   (4) Exploitation of higher echelon successes.

b. __Sniper Employment__. The rapid movement during the initial stages of mechanized infantry attack limits active sniper employment in support of the tank-tractor column. Sniper support, therefore, will be limited to periods when the armored grouping is halted and to the infantry attack subsequent to dismounting.

   (1) __Employment During Movement__. Sniper teams are normally embarked in the lead, center, and rear amphibious tractors. During voluntary administrative halts or involuntary halts due to enemy action, sniper teams debark immediately and assume firing and observation positions.

   (2) __Employment After Dismounting__. The infantry unit commander designates an area forward of the line of departure in which attacking platoons dismount from tractors to continue the attack on foot. The dismount area should be a concealed or covered area, if possible, and it should be located as far forward as the terrain and the enemy situation permit. Immediately after dismounting, sniper teams are deployed, if time permits, in the immediate area to provide long-range observation and precision long-range support fires if enemy targets appear. Upon commencement of the attack on foot, employment is in accordance with conventional offensive tactics.

c. __Consolidation__. After seizing the objective, the attacking force consolidates the position. Sniper teams are positioned to augment the observation and support fire capability of the tanks and amphibious tractors which are positioned for flank and rear security. As in the tank-infantry attack, the availability of firepower on the objective may permit rifle units to be withdrawn to covered positions for reorganization. Close-in protection requirements for the tanks and amphibious tractors are met in part by retention of sniper teams on the objective.

12. <u>Attack of Fortified Areas</u>. Fortifications provide a base for offensive operations or a series of strong defensive positions for the protection of vital areas. Fortified works of some nature are invariably constructed when military forces have a defensive mission. Depending upon the time and resources available for their construction, they range in complexity from simple, hastily prepared log or earth bunkers constructed from locally available materials, to permanent concrete and steel emplacements with fixed embrasures or steel turrets, intricate underground passages, and elaborate troop quarters. The fluidity of modern war normally limits opposing forces to the use of field fortifications constructed from locally available materials. These may include fortified weapons emplacements or bunkers, protected shelters, reinforced natural or constructed caves, entrenchments, and obstacles. Normally, emplacements and bunkers are mutually supporting and disposed in width and depth. The precision fire and observation capabilities of the sniper team are considered invaluable in the attack of a fortified area.

a. <u>Special Considerations</u>. The attack of a fortified area is usually difficult and requires special considerations. The enemy's cover, prepared obstacles, defensive fire plan, and carefully prepared counterattack plans give him definite advantages. For these reasons, fortified areas are usually bypassed by the main force and contained by a minimum force. There are, however, certain disadvantages in defending a fortified area which the sniper must exploit whenever possible. These weaknesses include:

(1) <u>Lack of Mobility</u>. Bunkers and emplacements cannot be relocated or altered to meet new threats or changing situations.

(2) <u>Openings</u>. Emplacements are weakest near embrasures, air vents, observation posts, and doorways. These various openings are vulnerable to precision fire.

(3) <u>Lack of Visibility</u>. A single embrasure in an emplacement can cover only a small sector of observation and fire. Lack of visibility makes one emplacement depend on another for support. The neutralization of each emplacement makes the defense progressively less effective.

b. <u>Sniper Team Employment</u>. Snipers employed in an attack on a fortified position are normally assigned the primary mission of delivering precision fire into observation posts, embrasures, and at exposed personnel. Targets are engaged selectively to ensure systematic reduction of the enemy's defenses through destruction of his mutual support capability. Penetration is the usual form of maneuver for attacks on fortified positions and snipers are employed accordingly.

c. <u>Position Requirements</u>. The sniper team establishes a position as close as possible to the area to be penetrated. Positions should be on the flanks of the zone of action. This permits continuous fire support not only for the assault units, but for adjacent units as well.

d. <u>Fire Support Plan</u>. The fire support plan of the infantry unit commander assigns specific tasks to snipers based on intelligence of:

(1) Exact locations and extent of individual fortifications.
(2) Locations and numbers of embrasures, fields of fire, and types of weapons therein.
(3) Locations of entrances, exits, and air vents in each emplacement.
(4) Directions of fire and types of fixed weapons.
(5) Extent of underground fortifications.
(6) Locations of natural and artificial obstacles.
(7) Locations of weak spots in the defense.

e. Consolodation. Upon capture of the objective, sniper teams displace forward to new positions from which to support a continuation of the attack or assist in repelling counterattacks.

13. Attack of Built-Up Areas

a. General

(1) Attack. The attack of a built-up area is divided into three phases:

(a) Phase I. Phase I is designed to isolate the battle area by seizing terrain features which dominate the approaches to it. Snipers deliver long-range precision fire at targets of opportunity.

(b) Phase II. Phase II consists of the advance to the built-up area and the seizure of a foothold on its edge. It is during this period that snipers displace forward and assume their initial positions from which to support continuation of the attack.

(c) Phase III. Phase III consists of the advance through the built-up area in accordance with the plan of attack.

(2) Special Considerations. The nature of attack on built-up areas may vary from one of complete destruction to a requirement for capture without major damage. In the former case, artillery will play the major role, and the sniper will be mainly involved only in consolidation. When the area is to be preserved, however, sniper fire will play a very significant part in the advance and special consideration must be given to the factors of control and terrain as they affect the employment.

b. Control. The advance through a built-up area will frequently consist of many separate and apparently independent actions. Control becomes decentralized and communication efficiency is lowered because of radio failure due to surrounding structure. In this situation, the sniper teams must have a very clear picture of the scheme and progress of maneuver if they are to provide timely and effective support.

c. Terrain. The terrain of a built-up area is, of course, entirely artificial and radically different from that of any other type of operation.

(1) Observation Areas and Fields of Fire. Observation areas and fields of fire are clearly defined by streets and highways, but the surveillance problem is tremendously complicated by the possibly hundreds of rooftops, windows, and doorways, each of which is a separate and distinct point for observation.

(2) Cover and Concealment. Built-up areas offer excellent cover and concealment for both attackers and defenders. The defender has a decisive advantage, however, because the attacker must expose himself to move through the area. The sniper has a very distinct advantage because he does not necessarily have to move to the most advance line. He may occupy a higher position to the rear or flanks some distance away from his unit.

(3) Avenues of Approach. The best avenues of approach are the building interiors since movement through the streets is so easily detected. Snipers, whether attacking or defending in built-up areas, must learn every possible avenue of approach in their areas of operation.

d. Sniper Employment.

(1) Assignment. Sniper teams should operate in each zone of action, moving with and supporting the infantry units. They should operate at sufficient distance from the riflemen to keep from getting involved in fire fights but close enough to kill more distant targets which should threaten the advance. Some sniper teams should operate fully independent of the infantry on missions of search for targets of opportunity and particularly for enemy snipers.

(2) Positions

(a) Mutual Support. In built-up areas, it is desirable that team members operate from separate positions. Detection of two men in close proximity is very probable, considering the wealth of positions from which the enemy may be observing. The snipers should locate themselves where they can provide mutual support.

(b) Camouflage and Cover. Fields of fire are obvious to the enemy, and he will be well aware of likely sniper locations. Camouflage and cover under such circumstances are very difficult, but the resourceful sniper will find ways to remain unnoticed.

1 Firing From Windows. When firing from a window, the sniper should, if possible, fire from a position back in the room. The sound will be muffled and the muzzle flash will not be noticed. If he must show his rifle or part of his body to make the shot, he should abandon that position after firing.

2 Loopholes. Instead of firing through windows or doorways, the sniper can gouge out of the wall a funnel shaped hole with the large end at the room's interior. Such a hole is inconspicuous, a poor target, and allows a considerable sector of fire.

<u>3</u> <u>Missed Shots</u>. The sniper should always abandon a position from which he has fired two or three misses. His detection is almost certain.

<u>4</u> <u>Traffic</u>. The sniper's position should never be subject to traffic or other personnel, regardless of how well the sniper is hidden. Traffic will invite observation, and the sniper may be detected by optical devises. He must abandon the position rather than risk detection.

14. <u>River-Crossing Operations</u>

a. <u>General</u>

(1) The purpose of a river-crossing operation is to move an attacking force rapidly across a river obstacle so that it may continue its attack to seize assigned objectives. Sniper teams, by virtue of their observation and precision fire capability, are uniquely adaptable to the initial stages of the river crossing.

(2) There are two types of river crossings: hasty and deliberate.

(a) <u>Hasty Crossing</u>. A crossing is termed hasty when it can be conducted as a continuation of the attack, with a minimum loss of momentum, by the same large forces which executed the advance to the river line. It is characterized by speed, surprise, and minimum concentration of personnel and equipment.

(b) <u>Deliberate Crossing</u>. The deliberate crossing is characterized by some delay, more detailed preparation, and the employment of extensive and specialized crossing means.

b. <u>Concept</u>

(1) The effectiveness of the river as an obstacle is reduced through surprise and deception, speed of attack, and rapid buildup of combat power on the opposite shore.

(2) River crossings are normally made on wide fronts to facilitate dispersion, rapidity, and deception.

(3) When possible, assault units cross in helicopters and/or amphibious vehicles to seize deep objectives. When helicopters or amphibious tractors are not available, assault units cross in boats or by constructed bridges. In this case, they are assigned objectives close to the river.

c. <u>Sniper Employment</u>

(1) <u>General</u>. Snipers are employed effectively in general support prior to and during the crossing.

(a) <u>Prior to the Crossing</u>. Sniper teams assume positions across the total width of the crossing area with the primary mission of observation. All sightings of enemy activity are immediately reported to higher authority.

(b) <u>During the Crossing</u>. During the crossing, sniper teams support the crossing by observation and suppression of the enemy's observation and fire. The precision fire capability of the sniper team makes continuous fire support possible up to the time the landing is effected and the troops commence movement inland.

(2) <u>Planning</u>

(a) The sniper team leader should conduct a joint reconnaissance with the infantry unit commander to determine the number of sniper teams necessary to support the crossing. Snipers must be placed in position as early as possible, preferably during the reconnaissance stage. The time of sniper displacement across the river for support of the continuation of the attack must be preplanned. Generally, displacement commences immediately after the first troops have reached the opposite shore.

(b) In the event helicopters are used for deep assault, a priority of sniper need is established for helicopterborne and surface units. If sufficient sniper teams are available, they should be attached to both elements.

(3) <u>Crossing in Boats</u>. Boat crossings are generally made during periods of reduced visibility. Limited employment of snipers during boat crossings is possible providing a full moon exists or artificial illumination is utilized. When crossings are made during periods of reduced visibility, sniper fire support must be lifted earlier than usual to prevent accidental shooting of friendly troops.

(4) <u>Firing</u>. During the actual crossing, snipers hold their fire, to preserve secrecy, unless targets appear which threaten the operation.

15. <u>Patrolling</u>. A patrol is a detachment sent out from a unit to perform an assigned mission of reconnaissance or combat, or a combination of both. Patrolling is one of the surest means of establishing and maintaining security, gaining information, and contacting, harassing, or damaging the enemy. The effective employment of sniper teams with any size or type patrol is limited only by the terrain and the ingenuity and imagination of the patrol leader. The succeeding paragraphs are not intended as a complete source of information on all aspects of patrolling in general. It is essential that snipers aquire a thorough knowledge of all aspects of patrolling.

a. <u>Types of Patrols</u>. Patrols are classified by the type mission performed. The two general classifications are combat and reconnaissance. Their principle difference is in the action at the objective.

(1) <u>Reconnaissance Patrols</u>. Reconnaissance patrols collect or confirm imformation. They are organized into a reconnaissance element and a security element. The reconnaissance element reconnoiters or maintains surveillance over the objective.

The security element secures the objective, rallying point, gives early warning of enemy approach into the objective area, and protects the reconnaissance element. Missions that may be assigned a reconnaissance patrol include the following:

     (a) Point Reconnaissance. A reconnaissance conducted to collect information about a specific location or a small specified area, usually a known position or activity.

     (b) Area Reconnaissance. A reconnaissance conducted within an area defined by boundaries or other limiting features. The reconnaissance unit is given maximum freedom of action within the assigned area.

     (2) Combat Patrols. Combat patrols are organized to perform the following missions:

     (a) Raids. A raid patrol is a surprise attack for the purposes of destroying the enemy installations and equipment, killing enemy personnel, capturing enemy personnel and equipment, or liberating personnel. Raids are conducted by small forces which rely upon surprise and coordination for success. Raids are frequently conducted at night or in bad weather to enhance the factor of surprise.

     (b) Economy of Force Actions. Economy of force patrols establish roadblocks to slow enemy movement, seize key terrain to deny enemy access to an area, cover a withdrawing friendly force, and block enemy interference with larger friendly offensive actions. The economy of force action differs from a raid in that it retains its objective.

     (c) Security. When a security patrol detects the enemy, it functions as a raid patrol to capture or kill or, in the case of a large enemy force, the security patrol will provide delaying action. In a moving situation, security patrols screen flanks, areas, and routes. In a static situation, they prevent the enemy from infiltrating the area, detect and destroy infiltrators, and prevent surprise attack.

     (d) Contact. Contact patrols establish or maintain contact with the enemy or between friendly forces. A contact patrol is organized and armed according to knowledge of the enemy situation and the size of his forces in the area. The contact patrol must be capable of overcoming screening forces in order to contact main forces.

     (e) Ambush. Ambush patrols carry out surprise attacks from concealment against an enemy party which is moving or at a temporary halt. Enemy patrols, carrying parties, foot columns, trains, or vehicle convoys are some ambush objectives.

     (f) Search and Attack. A search and attack patrol is a patrol with the general mission of seeking out and attacking targets of opportunity. This patrol is a combination reconnaissance and combat patrol which searches for, and within its capability, engages targets when and where found. Engagement is by raid, ambush, or any form of attack suitable to the situation.

b. Sniper Employment in Patrols

(1) Reconnaissance Patrols Generally, only one sniper team is attached to a reconnaissance patrol when it is assigned a point reconnaissance. If the patrol has an area or zone reconnaissance or surveillance mission, two or more teams may be attached. The snipers normally remain with the security element to provide long-range protection for the reconnaissance element. If terrain conditions permit, the long-range accuracy of the sniper rifle permits the reconnaissance element to patrol farther away from the security element yet remain within effective support range. The comparatively slow rate of fire of the sniper rifle limits its practicality as an "all-around" weapon for use with the reconnaissance element. To prevent compromise of the reconnaissance team position, the sniper team fires only in self-defense or when ordered by the reconnaissance patrol leader. Normally, the only appropriate time to fire at a target of opportunity is when extraction or departure from the position is imminent and firing will not endanger the success of the patrol

(2) Combat Patrols

(a) Raids. The decision to employ snipers on a raid is influenced by the time of day the raid is to be conducted and the desired size of the patrol. If the raid is at night, the employment of snipers is impractical. When maximum firepower is essential and the size of the patrol must be limited, snipers may not be included. If patrol size permits and long-range precision fire is needed, sniper teams should be attached. The sniper team is normally attached to the security element. If appropriate, the sniper team may be attached to the support element to assist in providing long-range supporting fires. When attached to the security element, the sniper team assists in observing, in preventing enemy escape from the objective area, and in covering the withdrawal of the assault force to the rallying point. Upon withdrawal from the rallying point, the sniper team may be left behind for a short period to delay and harass enemy counteraction or pursuit.

(b) Economy of Force Patrols. The sniper team is ideally suited for retarding enemy movement by the application of precision, long-range fire from well-concealed positions. The enemy is taken under fire at the longest range practical under the existing wind, visibility, and terrain conditions. As the enemy nears, the sniper teams become increasingly selective and concentrate on killing leaders, radiomen, and crew-served weapons personnel.

(c) Security, Contact, and Search and Attack Patrols. The sniper teams move with the main bodies of these patrols. They are not used as points because of their inability to deliver volume of fire. If the patrol is taken under fire, the sniper team immediately assumes a firing position and attempts to locate the enemy with the aid of the rifle telescopic sight and binoculars. The sniper team continuously estimates the range to areas from which there is a likelihood of being ambushed. Sight settings are changed to correspond with the estimated range to expedite retaliation in the event they are fire upon.

(d) Ambush Patrols. Sniper teams are positioned in areas which afford observation and fields of fire into terrain features which might afford the enemy cover after the ambush has been initiated. To provide maximum coverage of the ambush site, sniper teams should be located at both ends of the ambush.

The long range of the sniper rifle enables the sniper team to position themselves away from the main body. The fires of the sniper team are coordinated into the fire plan. The sniper seeks leaders, radio operators, and crew-served weapons personnel as primary targets. If the enemy is mounted in trucks, every effort is made to kill the drivers of lead and end vehicles to block the road, prevent escape, and to create confusion. Upon cessation of fire, snipers may be retained in position long enough to cover withdrawal of the ambush unit.

16. <u>Extended Daylight Ambush</u>. An extended daylight ambush is an ambush conducted exclusively by snipers from preselected positions in areas where there is a likelihood of encountering the enemy. It is employed to isolate areas within the battle area by restricting enemy movement, to create fear and confusion among enemy troops, and to gain information. Additionally, it can be established near any positions from which friendly forces have recently withdrawn, to ambush enemy troops who may filter into our old positions to scavenge for useable trash or lost or discarded equipment.

a. <u>Selection of Ambush Areas</u>. Air observers, intelligence reports, and patrols are prime sources of information in determining advantageous locations for ambushes. Trails, river crossings, routes of communication, and main supply routes are considered likely areas in which to encounter the enemy. Generally, the ambush is established within the effective support range of artillery. If it is established in areas of heavy enemy activity and the routes to and from the sniper team positions are not conducive to rapid and concealed movement, infantry troops should accompany the snipers. The size of the supporting infantry element will be determined by degree of enemy activity anticipated. The fewer personnel involved, the less likely detection will be.

b. <u>Selection of Specific Ambush Site</u>. When the general area has been established, a specific position from which to fire must be selected. The position should possess the following characteristics:

    (1) Maximum observation of the objective area.
    (2) Fields of fire.
    (3) Covered routes of ingress.
    (4) Natural camouflage.
    (5) Cover.

c. <u>Sniper Employment</u>. When the decision has been made to employ sniper teams in a certain area, the sniper team coordinates the anticipated ambush with the unit to which attached or the unit which has tactical responsibility for the proposed area. Matters to be discussed and/or coordinated will include: coordinates of ambush sites, time and routes of departure and return, passwords and countersigns, radio frequencies and call signs, fire and infantry support matters, and time and routes of friendly patrols in the area.

    (1) <u>Preparation</u>. Prior to departure, the sniper team, after briefing, should make a detailed checklist of preparatory actions and follow it systematically to ensure full readiness.

    (2) <u>Departure</u>. Departure to an objective area should commence during the hours of darkness to ensure that the sniper team is in position prior to first light. Rigid patrol discipline is maintained enroute.

(3) <u>Arrival at Ambush Site</u>. Immediately upon arrival at the ambush site, the area must be thoroughly and quickly reconnoitered. Positions must be established and made comfortable, hasty fields of fire cleared, and foliage gathered for camouflage. The firing position must be made usable prior to daylight. Dirt excavated from the position is disposed of by placing it in sandbags which will be used as protection and a rest upon which the rifle will be steadied.

(4) <u>Requirements of a Firing Position</u>. The sniper team's firing positions are located to provide maximum coverage of the entire area, consistent with the team concepts of mutual support and alternating the firing and observing duties. Darkness will make selection of mutually supporting positions difficult; however, every effort must be made to prepare the position as thoroughly as possible, even under the most adverse conditions.

(5) <u>Conduct of the Mission</u>. At first light, both members of the sniper team observe. During the early morning and at dusk, the enemy has a tendency to become careless and will expose himself. Also, enemy activity will increase at first light. Range cards are prepared as quickly as possible after daybreak. Prominent terrain and manmade features are compared with the map as an aid in determining range. A determination is made, from observation of the terrain, of where the enemy is most likely to appear. Wind values are estimated and compared continuously throughout the day. This procedure expedites firing a shot when a target appears. It is absolutely essential that the sniper team remain alert but motionless during the day.

(a) <u>Shooting</u>. When a target appears, a determination must be made whether or not to fire. Only targets that are positively identified as the enemy and are clearly defined are fired upon. Indiscriminate firing at poorly defined targets only serves to compromise the security of the mission. The sniper aims in as the enemy appears and his partner observes the target. If the sniper misses, the observer "calls the shot" for his partner who adjusts his sights or "holds off" and shoots again if the target is still exposed. If large bodies of troops appear, an artillery mission should be called down upon them.

(b) <u>Evasive Action</u>. If the sniper team fires, a decision must be made whether to remain in position or move. At long ranges, it is difficult to determine the exact origin of a rifle shot; however, repeated shots disclose the position. After shooting, the sniper team must be particularly alert for enemy movement or unusual activity. If activity is oriented towards their position and a covered route of withdrawal is available, it is better to move. However, the original position should be held as long as possible if there is no unusual enemy activity after firing. Movement will increase the probability of detection.

(c) <u>Departure</u>. Departure from the sniper position is executed during darkness to avoid detection. Every effort is made to restore the position to its natural state so that the area may be used again.

17. **Helicopter Insertion**. Helicopters may be used to insert sniper teams into areas of operation when:

    a. The selected sniper position is located an excessive distance from friendly lines.

    b. The situation requires immediate employment.

    c. The route to a selected sniper position is unduly difficult or heavily saturated with the enemy.

    d. It is desired to create an adverse psychological effect upon the enemy by killing key personnel deep in enemy controlled areas.

    e. There is a requirement for sniper fire to support a helicopterborne assault or to secure terrain around a landing zone.

    f. Required for diversionary actions or in response to ambushes or friendly forces.

OPPORTUNITY FOR QUESTIONS AND COMMENTS

SUMMARY

1. **Reemphasize**. We have examined all the different phases of sniper employment in offensive operations. The commander decides how to use his snipers, but you advise the commander on how to use them most effectively.

2. **Remotivate**. The lives of your teams and of the units you support will depend on how effectively you have trained and employed your men. Proper use of the material in todays class will ensure that the morale as well as the lives of the enemy will be in unquestionable jeopardy.

SI0017
L, D, A

_____
(DATE)

DEFENSIVE EMPLOYMENT

DETAILED OUTLINE

INTRODUCTION

1. **Gain Attention.** The mission in the defense may be to deny a vital area to the enemy, to protect a flank, to contain an enemy force, or to affect maximum attrition and disorganization upon the enemy. The defense may be assumed for one or more of the following purposes:

    a. To allow development of more favorable conditions for undertaking the offense.

    b. To economize forces in one area in order to concentrate superior forces for decisive offensive action elsewhere.

    c. To permit the employment of nuclear weapons.

    d. To ensure the integrity of an objective seized during the attack.

2. **Motivate.** Although widely considered to be strictly an offensive fighter in combat, the sniper must, at times assume the defense. His mission does not change and therefore, he will be expected to continue to deliver precision fire on selected targets from concealed positions.

3. **State Purpose and Main Ideas**

    a. **Purpose.** The purpose of this period of instruction is to provide the student with the knowledge of sniper employment in defensive operations.

    b. **Main Ideas.** The main ideas to be discussed are the following:

(1)	Fundamentals of Defense	(6)	Perimeter Defense
(2)	General Outpost	(7)	Reverse Slope Defense
(3)	Combat Outpost	(8)	Defense of Built-Up Areas
(4)	Security	(9)	Defense of a River Line
(5)	Area Defense	(10)	Mobile Defense
		(11)	Retrograde Operations

4. **Learning Objectives.** Upon completion of this period of instruction, the student will be, without the aid of references, knowledgeable and proficient in describing the employment and use of snipers in the various types of defense.

TRANSITION. The defense is the employment of all means and methods available to prevent, resist, or destroy an enemy attack. The use of snipers in varied defensive situations, if properly employed, will effectively enhance or augment a unit's defensive fire plan.

BODY

1. Fundamentals of Defense. The defense of any position is planned, organized, and conducted by applying certain fundamentals. These fundamentals do not have equal influence nor are they equally emphasized at different levels of command. Further, they may not apply to the same extent in different situations. The commander of the unit to which snipers are attached decides the degree to which snipers will participate in the defense. After an analysis of the terrain, snipers submit recommendations to the unit commander on employment and positions.

   a. Utilization of Terrain. The sniper must always take maximum advantage of the terrain by occupying positions which offer good observation, fields of fire, concealment and cover, and which control enemy avenues of approach into the defensive position.

   b. Security. The sniper team must adopt security measures to offset the attacker's advantages of initiative and flexibility, and to cause him to attack under unfavorable conditions. Every conceivable measure is taken by the sniper team to avoid ground observation and surprise from any direction.

   c. Mutual Support. Sniper teams are positioned so they can coordinate surveillance and reinforce each other by fire.

   d. All-Around Defense. Snipers organize for defense in all directions by establishing a system of primary and supplementary positions.

   e. Defense in Depth. Snipers are positioned in depth throughout the defense to ensure sustained sniper fire support. Snipers positioned near the forward edge of the battle area (FEBA) are vulnerable to concentrated attacks because of their limited volume of fire.

   f. Coordinated Barrier Planning. Barrier planning includes considerations for the employment of a series of natural and artificial obstacles to restrict, delay, block, or stop the movement of enemy forces. Snipers cover obstacles by precision fire.

   g. Coordinated Fire Planning. All defensive fires are carefully planned and provide for the following:

      (1) Bringing the enemy under fire as soon as he comes within effective range.
      (2) Delivering increasingly heavier fires as the enemy approaches the battle area.
      (3) Breaking up the assault by fires immediately in from of the battle area.
      (4) Destroying the enemy or ejecting him by fires within the battle area should he succeed in penetrating it.

h. **Rate of Sniper Fire.** The rate of sniper fire does not increase or decrease as the enemy approaches. Specific targets such as officers, NCO's, and radio operators are systematically and deliberately destroyed without sacrificing accuracy for speed.

2. **Covering Force.** A covering force is normally established to provide security forward of the general outpost for a specific period to provide time for the preparation of defensive positions, to disorganize the attacking enemy forces, and to deceive the enemy as to the location of the battle area. Sniper teams are assigned to covering forces in strength to augment their fires.

3. **General Outpost.**

   a. **General.** The general outpost warns of enemy approach and provides time for the forward forces to prepare positions in the battle area. It covers the withdrawal of reconnaissance forces when they are operating to the front. It prevents observation of the battle area and delays the enemy advance.

   b. **Sniper Employment.** Snipers are assigned to general outposts to provide long-range precision fire to cause premature deployment of advancing enemy forces and to augment the outpost's observation capability.

4. **Combat Outpost.** The combat outpost is a security echelon consisting of a series of outguards covering the foreground of the positions of the regiment in the battle area. Its mission is similar to that of the general outposts. Additionally, the combat outpost provides target information for fire support agencies. The strength and composition of the combat outpost will vary, however, it should always be augmented by a minimum of one sniper team.

   a. **Organization.** The forces on the combat outposts are disposed laterally in a series of outguards varying in strength. The outguards are positioned near the topographical crest of terrain offering long-range observation and fields of fire, covering the avenues of approach into the battle area. Sniper teams are placed in positions which offer the best long-range fields of fire and observation into areas which are not covered by fire or observed by other outguards.

   b. **Withdrawal Plan.** When the general outpost withdraws, the combat outpost commander deploys patrols forward to gain and maintain contact with the enemy. The withdrawal plan is prepared and coordinated with the frontline units concerned. It also provides for an orderly withdrawal of the outguards on predetermined routes to successive delaying positions. The plan provides for extensive employment of snipers during this period to cover the withdrawal.

5. **Security**

   a. **Local Security.** Local security consists of sentinel posts, patrols, and listening posts. Sniper teams may be assigned any task involving local security during daylight hours.

   b. **Flank Security.** Exposed flanks are secured by locating reserves to block principal avenues of approach. Sniper teams enhance all-around security by providing long-range observation and precision fires.

5. **Area Defense.** The area defense is a relatively compact defense in its basic form and is characterized by a strongly held forward defense area. The basic fundamentals of defense as they apply to the sniper are applicable to all the variations of the area defense. When sniper teams are attached to any unit to augment the unit's defensive posture, their employment is directed by the commander's plan of defense. The unit commander in making his reconnaissance, designates sniper positions which will support his defensive scheme. He considers the terrain, the capability of the rifle, the elements of a desirable sniper position, and the mission to be assigned the sniper.

    a. **Sniper Assignment.** The sniper team may be assigned a role with the forward elements or with the reverse element. The requirements of a good position generally favor assignment to the reserve unit.

        (1) **Forward Elements.** If the requirement to provide adequate protection of the FEBA is paramount; sniper teams are positioned slightly to the rear of the frontline rifle units to avoid enemy fire directed at the frontline units. The sniper team is assigned the responsibility for defending critical avenues of approach and for firing at targets of opportunity.

        (2) **Reserve Elements.** The unit commander considers the retention and positioning of a reserve, consistent with the requirement for adequate forces to defend the FEBA. The sniper team, when assigned to the reserve unit, supports the FEBA units by fire, protects key terrain features in the rear, and controls the most dangerous approaches through the battle area. The sniper team's observation capability is ideally suited for assignment to security and surveillance missions with the reserve unit.

    b. **Fire Support Plan.** Sniper fires are planned and coordinated along with the fires of all other organic and supporting arms. Initially, the sniper team subjects the enemy to long-range precision fire. If the enemy succeeds in penetrating the area, the sniper shifts his fire to targets of opportunity within friendly positions to help contain the penetration and support the counterattack. Fires are coordinated with adjacent sniper units to ensure overlapping fields of observation and fire.

7. **Perimeter Defense.** Since this type defense has many of the characteristics of the other area defense variations, only the peculiarities of the perimeter defense are discussed in this paragraph.

    a. **Sniper Employment.** Maximum emphasis is placed upon mutual support within the perimeter. The generally circular trace of the FEBA makes it difficult to employ snipers as a team. The team may be split to gain increased coverage of the area or to cover several critical avenues of approach into the perimeter.

    b. **Sniper Position.** Ideally, the sniper team is positioned on rising terrain near the center of the perimeter providing the position provides all-around observation of avenues of approach and good fields of fire. The position should also be near the reserve element's position to facilitate support of the reserve unit in the event of a counterattack.

8. Reverse Slope Defense

a. General. A reverse slope defense is one organized on the portion of a terrain feature that is masked by a crest, from enemy direct fire and ground observation from the front. All or any part of the forces on the FEBA may be on the reverse slope, depending on the terrain in the area to be defended. A successful reverse slope defense depends on control of the crest by fire or physical occupation.

b. Sniper Employment. The sniper team is positioned with the security group on or just forward of the topographical crest to provide long-range fire and observation. If a position on the topographical crest is not available, the sniper team should be located with the reserve unit on the next high ground to the rear of the FEBA from which they can support the frontline platoons by fire.

9. Defense of Built-Up Areas.
The considerations which affect sniper employment in the defense of a built-up area are very similar to those affecting an attack of a built-up area. An awareness of what the enemy is committed to do or what he can be expected to do in any given situation will facilitate the sniper's task.

a. Sniper Employment. Snipers are preferably positioned in buildings of masonry construction which offer the best long-range fields of fire and all-around observation. They are assigned various missions which include:

    (1) Countersniper fire,
    (2) Firing at targets of opportunity,
    (3) Denying the enemy access to certain areas or avenues of approach,
    (4) Providing fire support over barricades and obstacles,
    (5) Surveillance of the flanks and rear areas,
    (6) Supporting counterattacks, and
    (7) Prevention of enemy observation.

b. Sniper Position. The ideal sniper position is not necessarily located in close proximity to the frontlines. Buildings bordering both sides of a street minimize the effects of wind on the trajectory of the bullet and permit the establishment of positions farther away from the frontlines. Positions in inconspicuous masonry buildings which afford a field of fire, observation, and routes of ingress and egress are ideal. Alternate or supplementary positions should also be established in built-up areas.

10. Defense of a River Line.
Rivers constitute obstacles to an attack and natural lines of resistance for defensive and delaying action. The natural characteristics of a river, i. e., flat, unobstructed field of fire and known distance, are exploited by the sniper team to increase its capabilities. The defense of a river line is conducted using the same fundamentals employed in other forms of defensive combat.

a. Reconnaissance. The defense of a river line requires thorough reconnaissance. The most probable crossing sites are determined so they can be defended in force. Main considerations are the banks, approaches to the banks, topography of adjacent terrain, and road nets on both sides of the river. An analysis of the foregoing will assist in determining probable enemy crossing sites and the best defensive positions.

b. <u>Sniper Employment</u>. Snipers are initially employed with the covering force which remains on the enemy's side of the river to maintain contact with the enemy. Every effort is made to harass, delay his advance, and determine assembly positions and his probable crossing sites. When forced to retire, sniper teams move to predetermined positions on the friendly side of the river and assume a defensive posture.

c. <u>Sniper Positions</u>. The sniper position should ideally be located as far above or below possible fording sites consistent with observation and fields of fire. If it is necessary to assume positions directly across from the possible fording site, the position should be located as far back from the river line as possible to avoid enemy preparatory fires.

11. <u>Mobile Defense</u>

a. <u>General</u>. The purpose of a mobile defense is to destroy an attacking enemy. Minimum size forces are positioned in forward areas to warn of attack and block or impede the enemy advance or canalize it into preselected killing zones along the avenues of approach. The capability of the sniper to deliver long-range precision fire is invaluable to a mobile defense. The considerations and fundamentals governing defensive employment in general are applicable in the mobile defense.

b. <u>Sniper Employment</u>. Snipers should be assigned to any size unit assigned the mission of establishing strong-points. A strongpoint is normally a defensive position organized by a battalion or company in the forward defense area. Its mission is to slow down, divert, repel, or destroy the advancing enemy. It provides information from which the location of the enemy's main attack, strength, and direction of advance can be determined. If a sufficient number of sniper teams are available, they should also be assigned to the strongpoint reserves to cover withdrawal of the strongpoint. The highly mobile characteristic of the sniper is particularly adaptable to performing indepentent harassing and observation missions for the strongpoint forces.

12. <u>Retrograde Operations</u>.

A retrograde movement is any movement of command to the rear or away from the enemy. It may be forced by the enemy or it may be a voluntary movement. Such movements are classified as withdrawals, retirements, or delaying actions.

a. <u>Purpose</u>. Retrograde movements are conducted to achieve one or more of the following purposes:

(1) Harrass and inflict punishment upon the enemy.

(2) Draw the enemy into an unfavorable position.

(3) Permit the use of elements of a force elsewhere.

(4) Avoid combat under undesirable conditions.

(5) Gain time without fighting a decisive engagement.

(6) To disengage from combat.

b. **Sniper Employment.** The foregoing considerations will apply to sniper employment in varying degrees. Essentially, however, snipers are assigned missions of supporting the action by delaying and inflicting casualties upon the enemy, observation, covering avenues of approach and obstacles by fire, harassing the enemy by causing him to prematurely deploy, and if the situation permits, directing artillery fire on large groups of the enemy. Provisions must be made for communications to facilitate control of the sniper's withdrawal and to call for fire if required.

OPPORTUNITY FOR QUESTIONS AND COMMENTS

SUMMARY

1. **Reemphasize.** During this period of instruction, we discussed that any defense was planned, organized and conducted by applying basic fundamentals. It is up to the sniper to submit recommendations to the unit commander on employment and positions after he analyzes the situation through the use of these fundamentals.

   We then discussed the different types of defenses and the employment and position of snipers in each. (Reemphasize specific ones if required.)

2. **Remotivate.** It will be up to you either as a team leader, platoon sergeant or platoon commander to make recommendations to the unit commander conducting the defense as to the proper employment and positioning of your team or a group of teams.

SI 0018
L, D, A

_____
(DATE)

CONSTRUCTION AND OCCUPATION OF HIDES

DETAILED OUTLINE

INTRODUCTION

1. **Gain Attention.** We already know that as snipers, we have to, among other things, be able to kill with one shot and observe and report on enemy activity. To be able to do this effectively, we must of course, ourselves be in killing range of the enemy's fire and also be subject to his observation. If however, we can stalk into and fire from, a well constructed, bullet proof and camouflaged position, then our chances of remaining undetected and therefore remaining highly effective for a much longer period are very greatly increased.

2. **Motivate.** A sniper's hide, therefore, although defensive in nature, is very much an offensive weapon.

3. **State Purpose and Main Ideas.**

   a. **Purpose.** The purpose of this period of instruction is to make the student aware of the proper selection, construction, and occupation of hides.

   b. **Main Ideas.** The main ideas to be discussed are the following:

      (1) Location of Hides,
      (2) Construction,
      (3) Concealment,
      (4) The Use of Buildings and Hides as Firing Positions, and
      (5) Types of Hides

4. **Learning Objectives.** Upon completion of this period of instruction, the student will:

   a. Be able to proficiently select, construct, and occupy a hide effectively.

   b. Know the basic types of hides and in what situations each can be best employed.

TRANSITION: Whenever possible, a sniper should work from a hide, since such a position affords a certain amount of free movement without the danger of detection and also protection from the weather and enemy fire.

BODY

1. <u>Location of Hides</u>. The general location of a hide is determined by the ground to be covered. Detailed location, however, requires careful recon-naissance, and likely positions may be found in such places as hedgerows, ruined buildings, rubbish heaps, edges of cuttings, etc...

    a. <u>Isolated Cover</u>. Isolated and conspicuous cover should be avoided and tree top positions, though maybe useful for observation, are unstable firing positions and difficult to occupy or vacate.

    b. <u>Urban Areas</u>. In urban areas, a study of large scale street maps, sewer plans, low level oblique photographs taken from helicopters and street photography by patrols coupled with previous patrol reports will assist in indicating the general areas suitable for possible hide locations. Once these have been established, further ground reconnaissance can be carried out to determine the approach route, entry and exit point and arcs of observation.

2. <u>Construction</u>

    a. <u>Materials.</u> Hides may be built of stone, brick, wood or turf, but it should be remembered that they will more often than not have to be constructed at night in close proximity to the enemy, so there will rarely be a chance to put up anything of an elaborate nature. Whatever materials are used, every effort is to be made to make a hide bullet broof. It is important to remember that there will rarely be an opportunity to view the hide from the front, so careful thought must be given to camouflage techniques.

    b. <u>Fighting Hole.</u> The most common type of hide may well be one made by enlarging an existing fighting hole. A variation of this is the "belly hide" which is similar, except for it's depth, to the enlarged fighting hole.

    c. <u>Permanent Hides.</u> The semi-permanent or permanent type of hide will usually only be practical in situations where defensive positions are being prepared and engineer help may be available.

    d. <u>Construction.</u> The construction of the hide should always begin with the pit so that if necessary, the sniper has a position from which he can fight. Spoil must be carefully hidden, and may even have to be carried some distance away in sandbags.

    e. <u>Loopholes.</u> The construction of loopholes requires care and practice to ensure that they afford adequate cover of the field of fire. They should be constructed so that they are wide at the back and narrow in the front, though not so narrow that observation is restricted.

    f. <u>Comfort.</u> Within reason, a sniper and his partner should be as comfor-table as possible in a hide. They must have sufficient room to shoot and to observe with ease. Adequate headroom is essential, particularly since the sniper's head will be higher with the sniper rifle than with the service rifle, because of the telescopic sight. It may be possible to construct some form of seat, but it should be remembered that if the hide is not big enough, the users will soon get cramped and the standard of their efficiency will drop.

3. Concealment

a. **Concealed Approaches**. It is essential that the natural appearance of the ground remains unaltered, and that any camouflage done is of the highest order. The sniper must also remember that though cover from view is cover from aimed fire, all concealment will be wasted if the sniper is observed as he enters the hide. It follows, therefore, that concealed approaches to the hide are an important consideration, and movement around it must be kept to a minimum. Efforts must be made to restrict entry to and exit from the hide prior to darkness. Track discipline must be rigidly enforced.

b. **Screens**. Any light shining from the rear of a hide through the front loophole may give the position away. It is necessary, therefore, to put a screen over the entrance to the hide, and also one over the loophole itself. The two screens must never be raised at the same time. Snipers must remember to lower the entrance screen as soon as they are in the hide and to lower the loophole screen before leaving it. These precautions will prevent light from shining directly through both openings.

c. **Loopholes**. Loopholes must be camouflaged using foliage or other material which blends with or is natural to the surroundings. Logically, anything not in keeping with the surroundings will be a source of suspicion to the enemy and hence a source of danger to the sniper.

d. **Urban Areas**. In urban areas, a secure and quiet approach with the minimum number of obstacles such as crumbling wall and barking dogs is required. When necessary, a diversion in the form of a vehicle or house search can be set up to allow the sniper the use of the cover of a vehicle to approach the area unseen and occupy the hide.

4. The Use Of Buildings And Hides As Fire Positions.

a. **Disadvantages**. Buildings can often offer good opportunities as sniping posts under static conditions. They suffer, however, the great disadvantage that they may be the object of attention from the enemy's heavier weapons. Isolated houses will probably be singled out even if a sniper using it has not been detected.

b. **Preparation**. Houses should be prepared for use in much the same way as other hides, similar precautions towards concealment being taken; loopholes being constructed and fire positions made.

c. **Outward Appearance**. Special care must be taken not to alter the outward appearance of the house by opening windows or doors that were found closed, or by drawing back curtains.

d. **Free Positions**. The actual fire positions must be well back in the shadow of the room against which the sniper might be silhouetted, must be a screened.

e. **Loopholes.** Loopholes may be holes in windows, shutters of the roof, preferably those that have been made by shells of other projectiles. If such loopholes have to be picked out of a wall, they must be made to look like war damage.

f. **Observation Rest.** Some form of rest for the firer and observer will have to be constructed in order to obtain the most accurate results. Furniture from the house, old mattresses, bedspreads and the like will serve the purpose admirably; if none of this material is available, sandbags may have to be used.

5. **Firing From Hides.**

a. **Fire Discipline.** Fire from a hide must be discreet and only undertaken at specific targets. Haphazard harassing fire will quickly lead to the enemy locating the hide and directing fire to it.

b. **Muzzle Flash.** At dusk and dawn, the flash from a shot can usually be clearly seen and care must be taken not to disclose the position of the hide when firing under such circumstances.

c. **Rifle Smoke.** On frosty mornings and damp days, there is a great danger of smoke from the rifle giving the position away. On such occasions, the sniper must keep as far back in the hide as possible.

d. **Dust.** When the surroundings are dry and dusty, the sniper must be careful not to cause too much dust to rise. It may be necessary to dampen the surroundings of the loophole and the hide when there is a danger of rising dust.

6. **Types of Hides.** A hide can take many forms. The type of operation or the battle situation coupled with the task the snipers are given, plus the time available, the terrain and above all, the ingenuity and inventiveness of the snipers, will decide how basic or elaborate the hide can or must be. In all situations, the type of hide will differ, but the net result is the same, the sniper can observe and fire without being detected.

a. **Belly Hide.** This type hide is best used in mobile situations or when the sniper doesn't plan to be in position for any extended period of time. Some of the advantages and disadvantages are:

(1) It is simple and can be quickly built.
(2) Good when the sniper is expected to be mobile, because many can be made.

Disadvantages

(1) It is uncomfortable and cannot be occupied for long periods of time.

b. **Enlarged Fire Trench Hide.** This type hide is nothing more than an enlarged fighting hole with advantages being:

(1) Able to maintain a low silhouette,
(2) Simple to construct,
(3) Can be occupied by both sniper and observer,
(4) It can be occupied for longer periods of time with some degree of comfort.

<u>Disadvantages</u>

    (1) It is not easily entered or exited from.
    (2) There is no overhead cover when in firing position.

  c. <u>Semi-Permanent Hide</u>. This type hide resembles a fortified bunker and should always be used if time circumstances permit. The advantages are:

    (1) Can be occupied for long periods of time,
    (2) Gives protection from fire and shrapnel,
    (3) Enables movement for fire and observation,
    (4) Provides some comfort.

<u>Disadvantages</u>

    (1) Takes time to construct
    (2) Equipment such as picks, shovels, axes, etc. are needed for construction.

  d. <u>Shell Holes</u>. Building a hide in a shell hole saves a lot of digging, but needs plenty of wood and rope to secure the sides. Drainage is the main disadvantage of occupying a shell hole as a hide.

  e. <u>Tree Hides</u>. In selecting trees for hides, use trees that have a good deep root such as oak, chestnut, hickory. During heavy winds, these trees tend to remain steady better than a pine which has surface roots and sways quite a bit in a breeze. A large tree should be used that is back from the wood line. This may limit your field of view, but it will better cover you from view.

<u>OPPORTUNITY FOR QUESTIONS AND COMMENTS</u>

<u>SUMMARY</u>

1. <u>Reemphasize</u>. During this period of instruction, we covered the complete construction of hides and locating the best area for this construction. It should be stressed that the sniper should use his own imagination and initiative while constructing his hide. In conclusion, we discussed the various types of hides, advantages and disadvantages and in what situation a particular hide could be best used.

2. <u>Remotivate</u>. The type of hide you build will depend on a great many things. Time, Terrain, Type of Operation, Enemy Situation, and Weapons, so always construct a good defensive hide. It will keep you effective and keep you alive.

UNITED STATES MARINE CORPS
SCOUT/SNIPER INSTRUCTOR SCHOOL
Marksmanship Training Unit, Weapons Training Battalion
Marine Corps Development and Education Command
Quantico, Virginia 22134

SI-0019
L, D

_____
(DATE)

INTERNAL SECURITY EMPLOYMENT
TITLE

DETAILED OUTLINE

INTRODUCTION

1. **Gain Attention.** Imagine the Marine Corps suddenly committed to a peace keeping force in Beirut, Lebanon. Or, imagine being committed to preserve the peace and protect innocent lives and property in an urban enviroment such as Detroit or Watts during a "Big City" riot. It has happened before and could happen again! What is the role of the sniper? Is the sniper a valid weapon for employment in situations like this?

2. **Motivate.** The answer is most emphatically, yes!! We have only to look around us to see examples of how effective the sniper can be in this type of situation. Probably the best examples available to us are two recent British involvements: Aden and Northern Ireland. In both cases the sniper has played a significant role in the successful British peace keeping efforts. Remember, that one of the key principles of crowd control/peace keeping is the use of only minimum force. The sniper with his selective target indentification and engagement with that <u>one</u> well aimed shot is one of the best examples of the use of <u>minimum</u> force.

3. **State Purpose and Main Ideas**

   a. **Purpose.** To provide the student with the general knowledge needed to employ a sniper section in internal security type enviroments.

   b. **Main Ideas.** To explain the sniper's role in:

      (1) Urban guerrilla operations.
      (2) Hostage situations.

4. **Learning Objectives**. Upon completion of this period of instruction, the student will be able to:

    a. Employ a battalion sniper section in either sniper cordon, periphery O. P. or ambush operations.

    b. Construct and occupy an urban O. P.

    c. Obtain and use special equipment needed for internal security operations.

    d. Employ a battalion sniper section in a hostage situation.

    e. Select a hostage situation firing position taking into consideration the accuracy requirements and effects of glass on the bullet.

BODY

1. Urban Guerrilla Warfare

    a. **General**. The role of the sniper in an urban guerrilla enviroment is to dominate the area of operations by delivery of selective, aimed fire against **specific** targets as authorized by local commanders. Usually this authorization only comes when such targets are about to employ firearms or other lethal weapons against the peace keeping force or innocent civilians. The sniper's other role, and almost equally important, is the gathering and reporting of intelligence.

    b. **Tasks**. Within the above role, some specific tasks which may be assigned include:

        (1) When authorized by local commanders, engaging dissidents/ urban guerrillas when involved in hijacking, kidnapping, holding hostages, etc.

        (2) Engaging urban guerrilla snipers as opportunity targets or as part of a deliberate clearance operation.

        (3) Covertly occupying concealed positions to observe selected areas.

        (4) Recording and reporting all suspicious activity in the area of observation.

        (5) Assisting in coordinating the activities of other elements by taking advantage of hidden observation posts.

        (6) Providing protection for other elements of the peace keeping force, including fireman, repair crews, etc.

c. Limitations. In urban guerrilla operations there are several
limiting factors that snipers would not encounter in a conventional war:

(1) There is no FEBA and therefore no "No Mans Land" in which to
operate. Snipers can therefore expect to operate in entirely hostile
surroundings in most circumstances.

(2) The enemy is covert, perfectly camouflaged among and totally
indistinguishable from the everyday populace that surrounds him.

(3) In areas where confrontation beween peace keeping forces and
the urban guerrillas takes place, the guerrilla dominates the ground
entirely from the point of view of continued presence and observation.
Every yard of ground is known to them; it is ground of their own choosing.
Anything approximating a conventional stalk to and occupation of, a hide is
doomed to failure.

(4) Although the sniper is not subject to the same difficult
conditions as he is in conventional war, he is subject to other pressures.
These include not only legal and political restraints but also the
requirement to kill or wound without the motivational stimulus normally
associated with the battlefield.

(5) Normally in conventional war, the sniper needs no clearance to
fire his shot. In urban guerrilla warfare, the sniper must make every
effort possible to determine in each case the need to open fire and that
it constitutes reasonable/minimum force under circumstances.

d. Methods of Employment

(1) Sniper Cordons/Periphery O. P. 's

(a) The difficulties to be overcome in pl, ng snipers in
heavily populated, hostile areas and for them to remain undetected, are
considerable. It is not impossible, but it requires a high degree of
training, not only on the part of the snipers involved, but also of the
supporting troops.

(b) To overcome the difficulties of detection and to maintain
security during every day sniping operations, the aim should be to confuse
the enemy. The peace keeping forces are greatly helped by the fact that
most "trouble areas" are relatively small, usually not more than a few
hundred yards in dimension. All can be largely dominated by a considerable
number of carefully sited O. P.'s around their peripheries.

(c) The urban guerrilla intelligence network will eventually
establish the locations of the various O. P.'s. By constantly changing the
O. P.'s which are in current use it is impossible for the terrorist to
know exactly which are occupied. However, the areas to be covered by the
O.P.'s remain fairly constant and the coordination of arcs of fire and
observation must be controlled at a high level, usually battalion. It may
be delegated to company level for secific operations.

(d) The number of O.P.'s required to successfully cordon an
area is considerable. Hence, the difficulties of sustaining such an
operation over a protracted period in the same area should not be under-
estimated.

(2) Sniper Ambush

        (a) In cases where intelligence is forth coming that a target will be in a specific place at a specific time, a sniper ambush is frequently a better alternative than a more cumbersome cordon operation.

        (b) Close reconnaissance is easier than in normal operation as it can be carried out by the sniper as part of a normal patrol without raising any undue suspicion. The pricipal difficulty is getting the ambush party to its hide undetected. To place snipers in position undetected will require some form of a deception plan. This often takes the form of a routine search operation in at least platoon strength. During the course of the search the snipers position themselves in their hide. They remain in position when the remainder of the force withdraws. This tactic is especially effective when carried out at night.

        (c) Once in position the snipers must be prepared to remain for lengthy periods in the closest proximity to the enemy and their sympathizers.

        (d) Their security is tenuous at best. Most urban O.P.'s have "dead spots" and this combined with the fact that special ambush positions are frequently out of direct observation by other friendly forces makes them highly susceptible to attack, especially from guerrillas armed with explosives. The uncertainty about being observed on entry is a constant worry to the snipers. It can and does have a most disquieting effect on the sniper and underlines the need for highly trained men of stable character.

        (e) If the ambush position cannot be directly supported from a permanent position, a "back up" force must be placed at immediate notice to extract the snipers after the ambush or in the event of compromise. Normally it must be assumed that after the ambush, the snipers cannot make their exit without assistance. They will be surrounded by large, extremely hostile crowds, consequently the "back up" force must not only be close at hand but also sufficient in size.

c. Urban Sniping Hides/O.P.'s

    (1) Selecting the Location. The selection of hides and O.P. positions demand great care. The over-riding requirement of a hide/O.P. position is for it to dominate its area of responsibility.

        (a) When selecting a suitable location there is always a tendency to go for height. In an urban operation this can be mistake. The greater the height attained, the more the sniper has to look out over an area and away from his immediate surroundings. For example, if an O.P. were established on the 10th floor of an apartment building, to see a road beneath, the sniper would have to lean out of the window, which does little for the O.P.'s security. The locations of incidents that the sniper might have to deal with are largely unpredictable, but the ranges are usually relatively short. Consequently, an O.P. must aim to cover its immediate surroundings as well as middle and far distances. In residential areas this is rarely possible as O.P.'s are forced off ground floor level by passing pedestrians. But generally it is not advisable to go above the second floor, because to go higher greatly increases the dead space in front of the O.P. This is not a cardinal rule, however. Local conditions, such as being on a bus route, may force the sniper to go higher to avoid direct observation by passengers.

   (b) In view of this weakness in local defense of urban O.P.s, the principles of mutual support between O.P.s assumes even greater importance. The need for mutual support is another reason for coordination and planning to take place at battalion level.

   (c) The following are possible hide/O.P. locations:

    (1) Old, derelict buildings. Special attention should be paid to the possibility of encountering booby traps. One proven method of detecting guerrilla booby traps is to notice if the locals (especially children) move in and about the building freely.

    (2) Occupied houses. After careful observation of the inhabitants daily routine, snipers can move into occupied homes and establish hides/O.P.s in basements and attics. This method is used very successfully by the British in Northern Ireland.

    (3) Shops.

    (4) Schools and Churches. When using these as hide/O.P. locations, the snipers risk possible damage to what might already be strained public relations.

    (5) Factories, sheds, garages.

    (6) Basements and between floors in buildings. It is possible for the sniper team to locate themselves in these positions although there may be no window or readily usable firing port available. These locations require the sniper to remove bricks or stone without leaving any noticeable evidence outside of the building. To do this the sniper must carefully measure the width of the mortar around a selected brick/stone. He must then construct a frame exactly the size of the selected brick with the frame edges exactly the size of the surrounding mortar. He then carefully removes the brick from the wall and places it in his frame. The mortar is then crushed and glued to the frame so that it blends perfectly with the untouched mortar still in place. The brick/frame combination is then placed back into the wall. From the outside, nothing appears abnormal, while inside the sniper team has created an extremely difficult to detect firing port. Care must be taken however that when firing from this position dust does not get blown about by muzzle blast and that the brick/frame combination is immediately replaced. Another difficulty encountered with this position is that it offers a very restricted field of view.

    (7) Rural areas from which urban areas can be observed.

   (d) An ideal hide/O.P. should have the following characteristics:

    (1) A secure and quiet approach route. This should, if possible, be free of garbage cans, crumbling walls, barking dogs and other impediments.

    (2) A secure entry and exit point. The more obvious and easily accessible entry/exit points are not necessarily the best as their constant use during subsequent relief of sniper teams may more readily lead to compromise.

(3) Good arcs of observation. Restricted arcs are inevitable but the greater the arc the better.

(4) Security. These considerations have already been discussed above.

(5) Comfort. This is the lowest priority but never the less important. Uncomfortable observation and firing positions can only be maintained for short periods. If there is no adequate relief from observation, O.P.s can rarely remain effective for more than a few hours.

(2) <u>Manning the O.P./Hide</u>.

(a) Before moving into the hide/O.P. the snipers <u>**must**</u> have the following information:

(1) The exact nature of the mission (i.e. observe, shoot, etc.)

(2) The length of stay.

(3) The local situation.

(4) Procedure and timing for entry.

(5) Emergency evacuation procedures.

(6) Radio procedures.

(7) Movement of any friendly troops.

(8) Procedure and timing for exit.

(9) Any special equipment needed.

(b) The well-tried and understood principle of remaining back from windows and other apertures when in buildings has a marked effect on the manning of O.P.s/hides. The field of view from the back of a room through a window is limited. To enable a worthwhile area to be covered, two or even three men may have to observe at one time from different parts of the room.

(3) <u>Special Equipment for Urban Hides/O.P.</u> The following equipment may be necessary for construction of or use in the urban/O.P.

(a) Pliers. To cut wires.

(b) Glass Cutter. To remove glass from windows.

(c) Suction Cups. To aid in removing glass.

(d) Rubber Headed Hammers. To use in construction of the hide with minimal noise.

(e) Skeleton Keys. To open locked doors.

(f) Pry Bars. To open jammed doors and windows.

(g) Padlocks. To lock doors near hide/O.P. entry and exit points.

2. Hostage Situations

a. **General.** Snipers and commanding officers must appreciate that even a good, well placed shot **may** **not** **always** result in the instantaneous death of a terrorist. Even the best sniper when armed with the best weapon and bullet combination **cannot** **guarantee** the desired results. Even an instantly fatal shot **may** **not** prevent the death of a hostage when muscle spasms in the terrorist's body trigger his weapon. As a rule then, the sniper should only be employed when all other means of solving the situation have been exhausted.

b. **Accuracy Requirements**

(1) The USMC M40A1 Sniper Rifle is the finest combat sniper weapon in the world. When using the Lake City M118 Match 7.62 mm ammunition it will constantly group to within one minute of angle or one inch at one hundred yards.

(2) Keeping this in mind, consider the size of the target in a hostage situation. Doctors all agree that the only place on a man, where if struck with a bullet instantaneous death will occur, is the head. (Generally, the normal human being will live 8-10 seconds after being shot directly in the heart.) The entire head of a man is a relatively large target measuring approximately / inches in diameter. But in order to narrow the odds and be more positive of an instant killing shot the size of the target greatly reduces. The portion of the brain that controls all motor reflex actions is located directly behind the eyes and runs generally from ear lobe to ear lobe and is roughly two inches wide. In reality then, the size of the snipers target is two inches not seven inches.

(3) By applying the windage and elevation rule, it is easy to see then that the average USMC sniper cannot and should not attempt to deliver an instantly killing head shot beyond 200 yards. To require him to do so, asks him to do something the rifle and ammunition combination available to him cannot do.

c. **Position Selection.** Generally the selection of a firing position for a hostage situation is not much different from selecting a firing position for any other form of combat. The same guidelines and rules apply. Remember, the terrain and situation will dictate your choice of firing positions. However, there are several peculiar considerations the sniper must remember:

(1) Although the sniper should only be used as a last resort, he should be moved into his position as early as possible. This will enable him to precisely estimate his ranges, positively identify both the hostages and the terrorist and select alternate firing positions for use if the situation should change.

(2) If the situation should require firing through glass, the sniper should know two things:

(a) That when the M118 ammunition penetrates glass, in most cases the copper jacket is stripped off its lead core and fragments. These fragments will injure or kill should they hit either the hostage or the terrorist. The fragments show no standard pattern but randomly fly in a cone shaped pattern much like shot from a shotgun. The lead core of the bullet does continue to fly in a straight line. Even when the glass is angled to as much as 45° the lead core will not show any signs of deflection. (back 6 feet from the point of impact with the glass)

(b) That when the bullet impacts with the glass, the glass will shatter and explode back into the room. The angle of the bullet impacting with the glass has absolutely no bearing on the direction of the flight of the shattered glass. The shattered glass will always fly perpendicular to the pane of the glass.

d. Command and Control

(1) Once the decision has been made by the commander to employ the sniper, all command and control of his actions should pass to the sniper team leader. At no time should the sniper have to fire on someone's command. He should be given clearance to fire and then he and he alone should decide exactly when.

(2) If more than one sniper team is used to engage one or more hostages it is imperative that the rule above applies to all teams. But it will be necessary for the snipers to communicate with each other. The most reliable method of accomplishing this is to establish a "land line" or TA-312 phone loop much like a gun loop used in artillery battery firing positions. This enables all teams to communicate with all the others without confusion about frequencies, radio procedure, etc.

OPPORTUNITY FOR QUESTIONS AND COMMENTS

SUMMARY

1. Reemphasize. During this period of instruction we discussed urban guerrilla operations and hostage situations. In urban guerrilla operations we outlined the tasks and limitations common to all operations. We then discussed the two methods of employing snipers: (1) sniper cordons/periphery O.P.s and (2) sniper ambushes. We discussed selecting a position in an urban area and the most suitable locations for hides/O.P.s. Then we looked at how to man an O.P. and what special equipment you might need to construct and work in it.

In the discussion of hostage situations we examined the accuracy requirements and the position selection considerations common to all terrorist enviroments. We also discussed the command and control procedures for employing snipers in this type of role.

2. <u>Remotivate</u>. Remember, its not outside the realm of possibility that someday you or someone you've trained could find himself in this type of situation. At that time you'll take the test--let's hope we have no failures, because the political and social repercussions are too great a price to pay for one sniper who didn't prepare himself to put that one round on target.

UNITED STATES MARINE CORPS
SCOUT/SNIPER INSTRUCTOR SCHOOL
Marksmanship Training Unit, Weapons Training Battalion
Marine Corps Development and Education Command
Quantico, Virginia 22134

SI-1000
L, D

_____
(DATE)

BORESIGHTING
AND
ZEROING M40A1 AT 600 YDS
TITLE

DETAILED OUTLINE

1. State Purpose and Main Ideas.

    a. Purpose.

        (1) To make the student proficient in boresighting the M40A1.
        (2) To make the student capable of zeroing the M40A1 at 600 yds.

    b. Main Ideas.

        (1) Proper methods of boresighting.
        (2) Proper methods of zeroing.

BODY

1. Conduct of the Exercise.

    a. Boresighting.

        (1) With the bolt removed from the rifle, place the rifle on a solid support such as a sandbag, ammo can, etc.

        (2) Looking through the barrel, (at the chamber end) adjust the rifle until the desired aiming point is visible through the center of the bore.

        (3) Without disturbing the lay of the rifle, look through the telescope and observe the position of the crosshairs in relation to the aiming point.

        (4) If the crosshairs do not coincide with the aiming point, adjust the elevation and windage screw knobs until the crosshairs quarter the aiming point.

(5) Look again through the bore to insure the rifle has not moved.

(6) Raise the elevation on the sight 16½ minutes.

(7) The rifle is now ready to be fired to confirm the rifle and telescopic sight alignment.

b. **Zeroing.** After the boresighting, the student will be given his ammunition and will start the zeroing portion of the exercise.

(1) Assume a supported position.

(2) Load one (1) round and chamber.

(3) Fire and observe impact.

(4) Adjust windage and elevation to impact in the center of the target.

(5) Record all windage and elevation changes.

(6) Fire and observe impact.

(7) Adjust windage and elevation if needed to impact in the center of the target. Each student must correct windage and elevation if shots are not center of the target. Students will do so until a 600 yd line zero is established.

2. Conduct of the Pit Officer/NCO.

a. Insure that target, pasters, and spotters are available for each target.

b. Insure that all targets are spaced apart.

c. Insure that there are two (2) students on each target.

d. Contact the Conducting Officer/NCO when pits are ready to go, and that all students are on their targets.

e. On command from the Conducting Officer/NCO, all targets will run into the air.

f. After the boresighting has been done, the students on the line will start firing for a 600 yd line zero.

g. Once a shot has appeared on the target, the student will pull the target down into the pits. At that time he will put a spotter in the shot hole and run the target back into the air to show the student firing where he has hit. At that time the student firing will make his changes to impact in the center of the target. The student in the pits will not pull the target down again until he receives another shot on his target. Once the student does receive another shot on his target he will then pull the target down into the pits, taking out the old spotter and pasting up the old shot hole and placing the spotter in the new shot hole and running the target back into the air. This will carry on until the firing student is zeroed.

h. If a mark is called on a target the student will then pull his target into the pits, take out the old spotter and paste up his old shot hole and begin to look for a shot hole. If a shot hole cannot be found he will run up a clean target.

i. Once all students have completed firing the Conducting Officer/NCO will call a cease fire. At that time all targets will go into the pits. Once the firing line has been cleared a change in line and pit details will take place.

4. Course of Fire.

Yard Line	No. Rounds	Target	Score	Time
600	10	"A"	N/A	N/A

UNITED STATES MARINE CORPS
SCOUT/SNIPER INSTRUCTOR SCHOOL
Marksmanship Training Unit, Weapons Training Battalion
Marine Corps Development and Education Command
Quantico, Virginia 22134

SI-1001
L, D

_____
(DATE)

STATIONARY TARGETS
TITLE

DETAILED OUTLINE

1. Schedule.

Date	Time	Location / Yd Line	Total Rds Needed
05 Sept 78	1430-1630	R4 / 600 Yd Line	200 / 10
06 Sept 78	0700-1130	R4 / 300-600 Yds	900 / 15
07 Sept 78	0700-1130	R4 / 300-600 Yds	900 / 15
08 Sept 78	1300-1630	R4 / 300-600 Yds	900 / 15
11 Sept 78	0700-1130	R4 / 300-600 Yds	1200 / 20
12 Sept 78	1400-1630	R4 / 700-1000 Yds	800 / 10
19 Sept 78	1300-1630	R4 / 700-1000 Yds	800 / 10
20 Sept 78	0700-1130	R4 / 700-1000 Yds	800 / 10
09 Oct 78	0700-1130	R4 / 300-1000 Yds	700 / 05

2. Requirements.

a. Six (6) FBI silhouette targets for all the above dates.

b. In addition to normal combat equipment, each sniper team will be equipped with sniper rifle and binocular/M49 Spotting Scope.

c. Students' uniform as directed.

3. **State Purpose and Main Ideas.**

   a. **Purpose.** To make the student proficient in the proper holds for stationary targets at ranges of 100-800 yds with a 600 yd line zero. Also to make the student proficient in his individual movement to and from his firing positions.

   b. **Main Ideas.** The main ideas to be discussed are the following:

   (1) Proper methods of holds.
   (2) Proper methods of individual movement.

4. **Learning Objectives.** Upon completion of this period of instruction, the student will:

   a. Demonstrate holds for targets from 100-800 yds.

   b. Demonstrate techniques of the low crawl.

   c. Demonstrate techniques of movement of the bolt.

BODY

1. **Conduct of the Exercise.** The student will wear camouflage and move tactically during all firing exercises. Tactical movement on the Known Distance range will consist of a low crawl from behind the firing point to the firing point. The student will then assume a firing position and wait until his target appears. Once his target appears, the student will chamber a round, read his factors affecting bullet flight and impact, and engage his target with the proper holds. The student will fire 15 rounds, chambering one round at a time to insure proper bolt operations. After completion of firing, the student will low crawl back to his starting point. This conduct of fire at stationary targets will repeat itself at all known distance yard lines.

2. **Conduct of the Pit Officer/NCO.**

   a. To insure target paster and spotters are available for each target.

   b. To insure all targets are spaced apart.

   c. To insure that two (2) students are on a target.

   d. To contact the Conducting Officer/NCO when pits are ready for the firing exercise.

   e. On command from the Conducting Officer/NCO, all stationary targets will appear.

   f. Once a student has received an impact on his target, he will pull the target into the pits and put a corresponding spotter in the target, (white on black, black on white) and bring the target back up into the air and leave it there until he receives another shot. The student will do this until the total of rounds has been fired.

g. If the student hears the command "mark" and his target number, ne will pull the target into the pits and check for a shot hole.

h. Once all students have completed firing, the Conducting Officer/NCO will call a cease fire, clear and check all weapons and a line and pit detail change over.

3. Scoring.

a. The value of each shot will be determined by the formula $V = R/100$. Where V is the value of the shot and R is the range.
i. e. $V = 500/100$; $V = 5$.

b. A miss will be scored as a zero.

c. Passing score for a firing exercise will be 80% of the total points available.

UNITED STATES MARINE CORPS
SCOUT/SNIPER INSTRUCTOR SCHOOL
Marksmanship Training Unit, Weapons Training Battalion
Marine Corps Development and Education Command
Quantico, Virginia 22134

SI-1001
L, D

_____
(DATE)

STATIONARY TARGETS
UNDER
ARTIFICIAL ILLUMINATION
TITLE

DETAILED OUTLINE

SCHEDULE

Date	Times	Location / Yd Line	Rounds Needed
27 Sept 78	1900 - 2200	R4 / 200 - 500 Yd Line	900 / 100 Illum

REQUIREMENTS

1. Six (6) FBI silhouette targets.

2. 100 artificial illumination (Pop-ups) for the above dates.

3. In addition to normal combat equipment, each sniper will be equipped with sniper rifle and binoculars/M49 Spotting Scope.

1. State Purpose and Main Ideas.

    a. Purpose. To make the sniper proficient in the proper holds for stationary targets at ranges of 200-500 yds.

    b. Main Ideas. The main idea to be discussed is the following:

        (1) Proper methods of holds under artificial illumination.

2. Learning Objectives. Upon completion of this period of instruction the student will:

    a. Be able to use the proper holds for stationary targets at ranges of 200-500 yards at night.

    b. To make the sniper aware of the difficulties associated with engaging stationary targets under artificial illumination.

BODY

1. <u>Conduct of Firing Exercise</u>.

   a. See Reference Indicator SI-1001.

   b. Enclosure (1) to Reference Indicator SI-1001. On command from the Conducting Officer/NCO, the Pit Officer/NCO will set off the artificial illumination. The sniper must engage his target before the illumination goes out.

   c. All targets will be placed as in Reference Indicator SI-1002.

2. <u>Scoring</u>.

   a. The value of each hit will be determined by the formula $V = R/100$; V = value or hit, R = the range

   b. Misses will be scored as zero.

   c. Passing score for this firing exercise will be 80% of the total points available.

UNITED STATES MARINE CORPS
SCOUT/SNIPER INSTRUCTOR SCHOOL
Marksmanship Training Unit, Weapons Training Battalion
Marine Corps Development and Education Command
Quantico, Virginia 22134

SI-1002
L, D

_____
(DATE)

MOVING TARGETS
TITLE

DETAILED OUTLINE

SCHEDULE

Dates	Times	Location / Yd Line	Rounds Needed
21 Sept 78	1300 - 1600	R4 / 300 - 600 Yd Line	1200 / 20
22 Sept 78	0700 - 1130 1300 - 1630	R4 / 300 - 600 Yd Line	2400 / 40
25 Sept 78	0700 - 1130	R4 / 700 & 800 Yd Line	1000 / 25
26 Sept 78	0700 - 1130	R4 / 700 & 800 Yd Line	1000 / 25
27 Sept 78	0700 - 1100	R4 / 700 & 800 Yd Line	1000 / 25
29 Sept 78	0700 - 1130	R4 / 300 - 800 Yd Line	600 / 05
09 Oct 78	1300 - 1630	R4 / 300 - 800 Yd Line	600 / 05

REQUIREMENTS

1. Twelve (12) 12" FBI silhouette targets for all the above dates.

2. Seven (7) "A" type targets.

3. In addition to normal combat equipment, each sniper team will be equipped with sniper rifle and binoculars/M49 Spotting Scope.

4. Student uniforms as directed.

1. State Purpose and Main Ideas.

   a. Purpose. To make tne sniper determine tne proper leads necessary to hit a target walking or running at ranges of 100 to 800 yds.

   b. Main Ideas. The main ideas to be discussed are tne following:

      (1) Methods of leading a moving target.
      (2) Angle of target movement.
      (3) Speed of target.
      (4) Normal Leads.
      (5) Double Leads.

2. Learning Objectives. Upon completion of tnis period of instruction, the student will:

   a. Be able to understand tne proper metnods ot leading a target at ranges of 100-800 yds.

BODY

1. Conduct ot Firing Exercise.

   a. The student will wear camouflage and move tactically during the moving exercise. Tactical movement for tnis exercise will consist of a low crawl from behind the firing point to the firing point (approximately 10-15 yds) and back.

   b. Each sniper team will be given a block of targets that will be his firing position (approximately 35 ft long). Each sniper team must pick a firing position within his sector of fire and low crawl to his firing position.

   c. The student then must load five (5) rounds of ammunition into his sniper rifle and wait until a target appears in his sector of fire.

   d. Moving targets will appear on the far left sector first and far right sector second of tne snipers firing sector.

   e. When a target appears, the sniper's observer must tell the sniper where in the sector the target is, the wind element at the time of the sighting and any other element that may cause a error in a first round hit.

   f. After the sniper has engaged his target and it is a hit, the target will go down and move to tne far corners of tneir sector of fire and wait until all targets nave reac..ed tnis position. On command from the Pit Officer or NCO, all targets will be sent into the air and show their hits. The sniper's observer will then record his hold used and plot the impact of the shot.

g. If a target has reached the end of the sector and it has not been fired upon or hit, the student will bring the target into the pits. On command from the Pit Officer/NCO all targets will go into the air. A miss will be indicated by facing the back side of the target towards the firing line. On the command from the Pit Officer/NCO, all targets will be taken back into the pits, and again on command from the Pit Officer/ NCO, the next set of targets will come up and start to move from right to left in their sector of fire.

2. Conduct of Pit Officer/NCO.

a. To insure targets, pasters, and spotters are available for each target.

b. To insure all targets are spaced;

(1) Pit team 1 - 1-8
(2) Pit team 2 - 9-16
(3) Pit team 3 - 17-24
(4) Pit team 4 - 25-32
(5) Pit team 5 - 33-40
(6) Pit team 6 - 41-50

c. To insure that two (2) students are manning each moving targets.

d. To contact the Conducting Officer/NCO when pits are ready to start the firing exercise.

e. On command from the Conducting Officer/NCO, each student will raise a moving target at the far left sector of fire, (e. g. 1, 9, 17, 25, 33, 41) and start walking from left to right or to the end of the firing sector (e. g. 8, 16, 24, 32, 40, 50).

f. The Pit Officer/NCO must insure that all targets start at the far left sector of fire first. (e. g. targets 1, 9, 17, 25, 33, 41)

g. On command, raise all moving targets and walk to the end of the sector (e. g. 8, 16, 24, 32, 40, 50).

h. Insure that if a target is hit, the student pulls the target into the pits and walks to his far sector or end of his sector and waits. If a target is not engaged or the sniper fires and misses, insure that the target keeps moving until it reaches the end of the sector and then brought down into the pits.

i. On command from the Conducting Officer/NCO, all targets will appear to show the student their impact. If it is a miss, the back side of the moving target will appear. On command all targets will be taken back into the pits.

j. Again on command the next set of targets will start to move from right to left.

k. If the student hears the word "mark" in their sector of fire, he will pull the target down and look for a shot. If an impact hole can not be found, raise the moving target and walk to the end of the sector.

3. <u>Scoring</u>.

a. The value of each hit will be determined by the formula $V = R / 100$; V= value of the hit, R= the range
i. e. $V = 500/100 = 5$
         $V = 600/100 = 6$
         $V = 300/100 = 3$

b. Misses will be scored as zero.

c. Passing score for a firing exercise will be 80% of the total points available.

UNITED STATES MARINE CORPS
SCOUT/SNIPER INSTRUCTOR SCHOOL
Marksmanship Training Unit, Weapons Training Battalion
Marine Corps Development and Education Command
Quantico, Virginia 22134

SI-1002
L, D

_____
(DATE)

MOVING TARGETS
UNDER
ARTIFICIAL ILLUMINATION
TITLE

DETAILED OUTLINE

SCHEDULE

Dates	Times	Location / Yd Line	Rounds Needed	
03 Oct 78	1900 - 2200	R4 / 200 - 500 Yd Line	900 / 100	Illum
05 Oct 78	1900 - 2200	R4 / 200 - 500 Yd Line	900 / 100	Illum

REQUIREMENTS

1. Twelve (12) 12" FBI silhouette targets.

2. 100 Artificial Illumination (Pop-up) for all the above dates.

3. In addition to normal combat equipment, each sniper will be equipped with sniper rifle and binoculars/M49 Spotting Scope.

4. Student's uniform as directed.

1. <u>State Purpose and Main Ideas</u>.

    a. <u>Purpose</u>. To make the sniper determine the proper leads necessary to hit a target walking or running under artificial illumination at ranges of 200-500 yards.

    b. <u>Main Ideas</u>. The main ideas are as follows:

        (1) Methods of leading a target under artificial illumination.
        (2) Angle of target movement.
        (3) Speed of target.
        (4) Normal Leads.
        (5) Double Leads.

2. <u>Learning Objectives</u>. Upon completion of this period of instruction, the student will:

    a. Proper methods of leading a moving target at ranges of 200-500 yds.

<u>BODY</u>

1. <u>Conduct of Firing Exercise</u>.

    a. See Reference Indicator SI-1002.

    b. Enclosure (1) to Reference Indicator SI-1002. On command from the Conducting Officer/NCO, the Pit Officer/NCO will set off the artificial illumination.

    c. All targets will move as in Reference Indicator SI-1002.

UNITED STATES MARINE CORPS
SCOUT/SNIPER INSTRUCTOR SCHOOL
Marksmanship Training Unit, Weapons Training Battalion
Marine Corps Development and Education Command
Quantico, Virginia 22134

SI-1003
L, D

_____
(DATE)

UNKNOWN DISTANCE FIRING
TITLE

DETAILED OUTLINE

INTRODUCTION

1. Gain Attention. Range estimation for unknown distances is simply the process of determining the distances between two points. In most situation one of these points will be the observer's own position. The other point may be a target or prominent feature on the terrain.

2. Motivate. The ability to accurately determine range is a key skill which must be developed by the sniper to accomplish his mission.

3. State Purpose and Main Ideas.

   a. Purpose. The purpose of this firing exercise is to make the sniper student knowledgeable in range estimation and proficient in engaging targets at unknown ranges.

   b. Main Ideas. The main ideas are as follows.

      (1) Methods of estimating range.
      (2) Effects of sloping terrain.

4. Learning Objectives. Upon completion of this firing exercise the sniper will:

   a. Be able to determine range with the aid of maps, eye, telescopic sight, and range estimating methods found in the FMF 1-3B, and sight setting beyond 800 yds.

BODY

1. **Conduct of Firing Exercise.**

    a. Each team will be given five (5) targets at ranges that are unknown to the students.

    b. Each sniper team will be given ten (10) rounds of ammunition, five (5) rounds per student.

    c. Each sniper team will have five (5) targets to engage at ranges up to 1000 yds. Each target will have the sniper teams assigned number which will appear on five (5) targets at five unknown ranges.

    d. On command from the Conducting Officer/NCO the students will engage their targets witnout time limit.

    e. After each student has completed firing the targets will be taken in and scored.

    f. Targets will be scored as five (5) points per hit per round fired. Total points per team is 50 points. Passing score for this firing exercise will be 80% of the total points available per team.

UNITED STATES MARINE CORPS
SCOUT/SNIPER INSTRUCTOR SCHOOL
Marksmanship Training Unit, Weapons Training Battalion
Marine Corps Development and Education Command
Quantico, Virginia 22134

SI-1004
L, D

_____
(DATE)

INTERNAL SECURITY FIRING EXERCISE
TITLE

DETAILED OUTLINE

1. **Aim.** The aim is to develop in the sniper the ability to direct precision fire on a point target in a counter-terrorist/counter guerilla situation.

2. **Description.** This exercise will be conducted in two firing periods of ½ day each. Thirty rounds will be fired each day at modified DEA silhouette targets from ranges of 100, 200 and 300 yards. Firing will be done in the prone supported, barricade supported and sling supported positions.

3. **Reconnaisance by Conducting Officer/NCO.** The range used in this exercise must have at least five firing positions each on the 100, 200 and 300 yard lines. In addition the following criteria must be met:

   a. There must be communications between the line and the pits.

   b. Wind flags should be placed at each yard line.

   c. One team (2 men) must man each point used in the pits.

4. **Conduct of the Exercise.**

   a. **Briefing.** On the firing line the student will be briefed on the following:

   (1) Aim of the exercise.
   (2) Target and relay assignments.
   (3) Types of targets to be used.
   (4) Number of rounds to be used.
   (5) Time limits.
   (6) Standards to be achieved.

   b. After the briefing students are moved to the ready line or the pits as assigned.

c. On both days, ten rounds each will be fired from the 100, 200 and 300 yard lines at the modified DEA silhouette target. The target is modified by blackening an area one inch above and below the centerline of the eyes and extending the width of the head. Only shots hitting within this area will be scored as hits. Sights will be adjusted at each yard line; hold-offs will not be used.

d. On the first day each student will fire at a "single" target with all thirty rounds.

e. On the second day the first five rounds will be fired at double targets, i. e. the terrorist target will be half-covered by a "hostage" target. The second five rounds will be fired at multiple targets, i. e. the "terrorist" target will be partially covered by two "hostage" targets. Since only shots within the darkened core of the head will count as hits, care must be taken to cover part of that core on the "terrorist" target with the "hostage" target.

f. Additionally, on the second day, the firing of rounds from each yard line will be done as follows:

      (1) Rounds 1 & 2 from behind a "wall" barricade.
      (2) Rounds 3 & 4 from a "roof" platform.
      (3) Rounds 5 & 6 from behind a post.
      (4) Rounds 7 & 8 from a "window".
      (5) Rounds 9 & 10 from the sling in prone position.

5. <u>Standards</u>. A student is deemed to have failed if:

a. He misses seven or more times each day.

b. If he hits a hostage at any time.

UNITED STATES MARINE CORPS
SCOUT/SNIPER INSTRUCTOR SCHOOL
Marksmanship Training Unit, Weapons Training Battalion
Marine Corps Development and Education Command
Quantico, Virginia 22134

SI-1005
L, D

_____
(DATE)

MARKSMANSHIP TEST
TITLE

DETAILED OUTLINE

1. **State Purpose and Main Ideas.**

    a. **Purpose.** The purpose of this marksmanship test is to evaluate tne student on his ability to engage 25 designated targets at various ranges, scoring one point per hit with 80% accuracy.

    b. **Main Ideas.** The main ideas included in this test are as follows:

        (1) Engaging stationary targets at ranges from 300-800 yards.
        (2) Engaging moving targets at ranges from 300-800 yards.
        (3) Conduct of Line Officer/NCO.
        (4) Conduct of Pit Officer/NCO.
        (5) Test scoring.
        (6) Sniper Qualification Course.
        (7) Line commands.

2. **Learning Objectives.** Upon completion of this marksmanship test the student will be able to engage stationary and moving targets at ranges from 300-800 yards with 80% accuracy.

3. **Equipment Requirements.**

    a. Two AN/PRC-77 w/ batteries

    b. (1) Seven type "A" targets
       (2) Six FBI silhouette targets
       (3) Twelve 12" FBI silhouette targets

4. **Student Requirements.** The student will wear camouflage and move tactically during the marksmanship test. Tactical movement for this test will consist of a low crawl from behind the firing point (approximately 5-10 yds) to the sniper's firing position.

BODY

1. **Conduct of Engaging Stationary Targets at Ranges from 300-800 Yards.**

    a. Each team will be assigned a block of eight targets each block of which will be so designated with the left and right limits marked with a 6 X 6 cardboard target frames mounted in the two respective carriages. Thus, the right limit for one block of targets will also serve as the left limit of the succeeding block etc. The following targets will serve as left and right limits respectively: 1, 8, 15, 22, 29, 36, and 43. The stationary target will be mounted in the middle carriage of the block of eight targets.

    b. The first stage of fire at each yard line (300, 500, 600, 700 and 800) will be engaging stationary targets from the supported prone or Hawkins position with a 600 yd zero on the weapon. Command will be given from the center of the line to load three rounds. The sniper and partner will have three minutes in which to judge wind, light condition, proper elevation hold and fire three rounds with the target being pulled and marked after each shot. After the three minute time limit has expired all stationary targets will be pulled, cleared and will remain in the pits. There will not be a change over between sniper and observer until the sniper has engaged his moving targets.

2. **Conduct of Engaging Moving Targets at Ranges from 300-800 Yards.**

    a. Each team will remain at their respective firing point to engage their moving targets within the assigned block of eight targets. One of the two butt pullers will position himself on the left limit and the other on the right limit with the appropriate moving target.

    b. The second stage of fire at each yard line (300, 500, 600, 700, and 800) will be engaging moving targets. The command will be given from the center of the line to load two rounds. Once the entire line is ready a moving target will appear on the left limit of each block of targets moving left to right. The sniper and partner will have approximately 15 to 20 seconds (whatever amount of time it takes the student to walk from the left limit to the right limit) in which to fire one round. The next target will move from the right limit to the left and again the sniper and his partner will have 15 to 20 seconds in which to fire one round. The targets will not be run up to show each hit. It will be up to the observer to advise the sniper on where his rounds are impacting (high, low, left, right or center).

3. **Conduct of the Line Officer/NCO.**

    a. It will be up to the Line Officer/NCO to see that the entire test is run smoothly and safely.

b. He will be the deciding factor should any complications or differences arise.

c. He will be responsible for briefing the Pit Officer/NCO on the conduct of the test and any other major items that he can foresee that will aid him in controlling the conduct of the test.

d. He will insure that all commands are given clearly and precisely and that all students are allotted the same amount of time for firing.

e. He will insure the required amount of ammunition is present and the appropriate range is signed out.

f. He will be responsible for the police of all firing lines.

4. Conduct of the Pit Officer/NCO.

a. It will be his responsibility to see that the test is run smoothly and safely for all individuals in the pits.

b. He will contact the Line Officer/NCO should any complications or differences arise in the pits.

c. He will be responsible for briefing all students as to the conduct of fire and how and when the particular types of targets will be run.

d. He will insure that all commands are given clearly and precisely and that the individuals on each block of targets record the number of hits received during the moving target stage and those numbers passed to the Line Officer/NCO.

e. He will insure that the required amount of targets are readily available and that each block of targets is correctly implaced and properly manned.

f. He will be directly responsible for the police of the pit area.

5. Test Scoring. Scoring will be conducted on the firing line as well as in the pits. Each student will fire 25 rounds at an assortment of stationary and moving targets from 300 to 800 yards. Each round will be valued at one point with a total point value of 25 points. Passing score for the test will be 80% of the total hits available. A miss will be scored as zero.

6. <u>SNIPER QUALIFICATION COURSE</u>

<u>Stage</u>	<u>Yd Line</u>	<u>*Tgt Type</u>	<u>Tgts Fired</u>	<u>No Rds</u>	<u>*Pos</u>	<u>Time Limit</u>	<u>Score</u>
1	300	S	3	3	SP/H	3 min	
2	300	M	2	2	SP/H	15/30 sec	
3	500	S	3	3	SP/H	3 min	
4	500	M	2	2	SP/H	15/30 sec	
5	600	S	3	3	SP/H	3 min	
6	600	M	2	2	SP/H	15/30 sec	
7	700	S	3	3	SP/H	3 min	
8	700	M	2	2	SP/H	15/30 sec	
9	800	S	3	3	SP/H	3 min	
10	800	M	2	2	SP/H	15/30 sec	

*   S - Stationary targets
   M - Moving targets
 SP - Supported prone
  H - Hawkins position

7. <u>Line Commands</u>.

    a. <u>Stationary Targets</u>. "Your (stage) of the Sniper Qualification
Course is your (yd line) stationary targets. Firing three (3) rounds in
a total time limit of three (3) minutes. With three (3) rounds load and
be ready. Is the firing line ready, ready on the right, ready on the left,
all ready on the firing line. You may engage your targets when they
<u>appear</u>."

    b. <u>Moving Targets</u>. Your (stage) of the Sniper Qualification Course
is your (yd line) moving targets, engaging two (2) targets, firing one (1)
round at each target with a 15 second time limit per target. With two (2)
rounds load and be ready. Is the firing line ready, ready on the right,
ready on the left, all ready on the firing line. You may engage your
targets when they <u>appear</u>."

UNITED STATES MARINE CORPS
SCOUT/SNIPER INSTRUCTOR SCHOOL
Marksmanship Training Unit, Weapons Training Battalion
Marine Corps Development and Education Command
Quantico, Virginia 22134

SI-1010
L, D
_____
(DATE)

HIDE CONSTRUCTION EXERCISE
(HIDE AND SEEK)

DETAILED OUTLINE

1. **Aim.** The aim is to develop in the sniper the ability to construct a semi-permanent firing position which will provide him with adequate cover, concealment and fields of fire.

2. **Description.** Having studied the terrain, the sniper will choose a position for a hide, within the limits provided and construct a hide which will not be visible to an observer who can be as little as 25 yards away.

3. **Reconnaissance by the Conducting Officer/NCO.** The area used for the hide construction exercise must be chosen with care. At a minimum, half again as many more possible locations must be available as there are teams to use those areas. The student must be given adequate time to construct his position before the observer begins his search. Boundry limits for the exercise must be clearly marked by natural or man made features.

4. **Conduct of the Exercise**

   a. **Briefing.** In the area where the exercise will be conducted, the student will be briefed on the following:

      (1) Aim.
      (2) Boundries.
      (3) Time limit (usually nine hours).
      (4) Direction from which the observer will be watching and moving.

   b. After the briefing, the teams will move into the exercise area to begin preparations.

   c. During the seek portion of the exercise, the observer will be in radio contact with two walkers. The walkers will be positioned in the exercise area and will move to detected hide positions when directed by the observer.

d. Along with the observer, the assistant instructor (AI) for the problem will move through area holding up cue cards, which the concealed students must observe and copy down.

e. If the students have not been observed, while the observer searches from 600 yds away, using an M49 scope and binoculars, the observer will move successively to distances 500 yds away, 300 yds, 100 yds and 25 yds. At all these distances he will have unlimited time to search for the students, except at 25 yds. At a distance of 25 yds the observer will have only one minute to search.

f. <u>Standards</u>. The student is deemed to have failed if:

    (1) He is detected at any time
    (2) He has not correctly copied all the cue cards.

Notes

Other Books Available From Desert Publications